AMERICAN ♦ CLASSICS

CADILLAC ELDORADO

James W. Howell and Jeanna Swanson Howell

Retired GM Designer Dave Holls and his 1932 Cadillac.

*Dedicated to
David R. Holls
—a designer's designer who respects the heritage and history of his craft.
—a designer who honored the Eldorado tradition with his work and genius from its
birth in 1953 to the creation of the revolutionary Cadillac Eldorado of 1992.*

First published in 1994 by Motorbooks International Publishers & Wholesalers, PO Box 2, 729 Prospect Avenue, Osceola, WI 54020 USA

© James W. Howell, 1994

All rights reserved. With the exception of quoting brief passages for the purposes of review no part of this publication may be reproduced without prior written permission from the Publisher

Motorbooks International is a certified trademark, registered with the United States Patent Office

The information in this book is true and complete to the best of our knowledge. All recommendations are made without any guarantee on the part of the author or Publisher, who also disclaim any liability incurred in connection with the use of this data or specific details

We recognize that some words, model names and designations, for example, mentioned herein are the property of the trademark holder. We use them for identification purposes only. This is not an official publication

Motorbooks International books are also available at discounts in bulk quantity for industrial or sales-promotional use. For details write to Special Sales Manager at the Publisher's address

Library of Congress Cataloging-in-Publication Data

Howell, James.
 Cadillac Eldorado/James W. Howell and Jeanna Swanson Howell.
  p. cm.
  Includes index.
  ISBN 0-87938-879-X
  1. Eldorado automobile—History. I. Howell, Jeanna Swanson. II. Title.
 TL215.E39H68   1994
 629.222'2—dc20   94-23072

On the front cover: A stunning fifties triumvirate. These three Eldorados are from California collector Leo Chu's garage. At left is a 1954, at right a 1957, and in the rear a 1959. *Dennis Adler*

On the title page: A "Cadillac V-12" rendering penned by Wayne Kady.

Printed and bound in the United States of America

A M E R I C A N ♦ C L A S S I C S

Contents

	Foreword *by Sergio Pininfarina*	4
	Preface	5
Chapter 1	**Background for a Legend**	7
Chapter 2	**1953: The First Cadillac Eldorado**	15
Chapter 3	**1954–1960: Fins and Elegance**	25
Chapter 4	**1961–1970: A Legend Reborn**	67
Chapter 5	**1971–1985: Showboat and Flagship**	99
Chapter 6	**1986–Present: Mistake and Masterpiece**	133
Chapter 7	**The Future of Cadillac Eldorado**	155
	Index	160

A M E R I C A N ♦ C L A S S I C S

Foreword

by Sergio Pininfarina

Sergio Pininfarina, president of Pininfarina, Turin, Italy. Many of the world's beautiful automobiles, including Ferrari, bear his company's signature logo.

For years and years I have been thinking of the Cadillac Eldorado as a myth.

This car excited my imagination and made me dream, as it did—I believe—for most people; for me it was the richest, most refined automobile in the U.S.A., and unsurpassed in its kind all over the world.

Also its name—Eldorado—contributed to its fabulous image of richness and happiness as this country did with the pioneers.

In the 50s this impression of richness reached a peak: the expression of the highest American standard of living of its time. In some cases, stylingwise, the Eldorado was exaggerated: for example, the heavy grille and bumpers, the tailfins, the elaborate sheet metal treatment.

In the 60s, the Eldorado was able to eliminate these excesses evolving towards simplified lines and decorations but continued to transmit an unchanged message of a super car. Not only, but it was able to create very distinctive styling solutions, like the shape of the roof and the design of the quarter light windows; its vertical rear lights became a classic—a typical distinctive element.

In the 80s the Eldorado badly suffered from a downsizing operation. In my opinion, it lost a part of its character and its image dangerously approached that of a common car.

I had the great pleasure of being asked to help in the design of the last Eldorado; I approached this problem with much respect. I tried to retain, in a modern way, the distinctive styling elements which had made this car great over the years, looking for simplicity of lines, harmony of proportions, majesty of the overall impression.

It is not up to me to say how much my contribution was important, but I think the present Eldorado is a great success, getting back to the grassroots of its image.

Of one thing I am particularly impressed: the refinement of every exterior and interior detail. This is a feature common with Pininfarina.

It certainly comes from the fact that those who have been working on this project—as it happened with the most successful Eldorados—show a real love for the automobile in its finest expression.

Sergio Pininfarina
May 23, 1994
Turin, Italy

Cadillac Eldorado Brougham "Jacqueline," a proposal by Pininfarina for Cadillac in 1961. Named for President Kennedy's popular wife, this design would have replaced that of the Cadillac Eldorado Broughams that were built by the Turin company in 1959 and 1960.

American Classics

Preface

Some Things to Consider Before Reading this Book

Today's advertising says, "Cadillac Eldorado, creating a higher standard." In this book we tell the story of the people who have upheld that standard for more than forty years. The Eldorado is the flagship of Cadillac Division—it is at the cutting edge of what an automobile can be. It is luxurious, sporty, personal, and elegant, but, more than anything, the Eldorado is *American*.

Each Cadillac Eldorado has been born of the creative minds of some of the world's top engineers and designers. When we write about cars, we like to tell the stories of how they were created, how they evolved out of philosophy and sketches to be denizens of the road. Each Eldorado was created much the same way, beginning with lines on paper, years before coming to life at the end of the production line.

Who Designed What?

People outside the industry are often mystified by how a car is designed and produced. Just a few months ago, an automobile enthusiast told us that the Eldorado of a particular year was designed by Harley Earl. We were dumbfounded. Although the man in charge bears the ultimate responsibility, the process of automotive design is not so straightforward.

As an example, let's consider one of the Eldorados from Harley Earl's years. Earl was the head of design at General Motors for decades. In fact, he instituted the discipline at GM. His methods are carried out around the world even today. It is true that every car manufactured during his tenure at GM had his mark on it. His influence was constant during that period. But he did *not* sit down and draw a design, then tell the engineers to go out and build it! What he did do was manage several staffs of designers in GM's various divisions. He critiqued their work. He told them what he wanted and what he did not want. He cultivated and appreciated good designers, rewarding them handsomely for their work. If someone displeased him, that person might suddenly find himself without a job.

When Cadillac Eldorado was just a glimmer in GM's eye. First announced in December 1929, during the 1930s Cadillac produced a prestigious special production automobile, the fabulous V-16. This picture shows one of several that made an introductory tour through Europe beginning in the summer of 1930. The trip included a visit to the ancestral home of the Cadillac family in France. From 1930 until its last production model in 1940, this fine automobile appeared in several body styles. (One unique incarnation was Cadillac's, indeed the industry's, first real show car, the V-16 Aerodynamic Coupe displayed at the Chicago World's Fair in 1934.) Not until 1953 did Cadillac produce yet another prestigious special production car, *the first Eldorado.*

To identify a particular car designed by one particular person is extremely rare. To identify the primary designer is a bit easier. We could only name a few such automobiles and their designers. Bob Scheelk was the primary designer of the 1957 Eldorado Brougham. Several designers worked with him, but his was the primary influence. Ned Nickles was the primary designer of the 1963 Buick Riviera. Again, there were other designers working with these masters in both cases.

In the case of the 1957 Brougham, Harley Earl was vice president of design at GM at the time. While his hands most likely never touched the car, ultimately the design was his responsibility because he had the power of veto and the force of his much-sought-after encouragement. As head of the Cadillac Studio, Ed Glowacke also held a great deal of responsibility for the car, but the primary source of ideas for that particular project was Bob Scheelk.

When Nickles designed the Riviera, he did it with Bill Mitchell's vision, help, and encouragement. In fact, when you know the whole story, you have to agree that the Riviera was Mitchell's baby, even though Nickles brought it into being. More importantly, Mitchell, who followed Earl as vice president of design, sold the car to GM's top management. You can see the influence of Earl and Mitchell in all the cars created during their respective tenures.

We bring this up because in reading automotive history, you might see apparent contradictions about who did what in developing a certain design. Obviously, time is a problem because we are usually talking about events that took place years, if not decades, ago. Do you have a clear idea of what you did at your job the first week of last month? And even if a designer can rely on memory, there are two other factors at work here:

First, when a design is successful, designers, being human beings, naturally tend to have warm memories of their work on that design and their contributions toward it. An unsuccessful design might tend to make one forget it was part of his or her career.

Second, the design of an automobile is such a complex affair that it is difficult to sort out the origin of various components. Considering this, one is reminded of the story of the ten blind men and the elephant. Each was able to touch only one part of the elephant and describe his experience from the standpoint of what he alone had experienced. For instance, one held the tail and described the elephant as being like a rope. Another touched a leg and described it as being like a tree trunk. You can see what we mean: the design process is complex, and any participant in that process can only describe what he or she has experienced. This latter point is most likely the ultimate reason for any apparent contradictions between two designers' recollections.

Finally, there is the obvious—but all too often forgotten—point that automobiles are not just the products of designers. Even before a car design leaves Design Staff at GM, it has undergone rigorous and extensive work by a host of talented modelers and body engineers. After this, production engineers and other professionals put in long hours bringing the design into reality. From other avenues come the contributions of engine, drivetrain, and suspension engineers—just to name a few. Management and marketing play an important role. An organization must be put together and maintained to produce the cars, and sales must be substantial to turn a profit for the corporation. In short, when you take delivery of a new car, it is the product of the genius and work of many people.

So, What's the Story Here?

Books on classic automobiles fit into many categories. Some will tell you whether or not a particular screw in a window molding should be chromed or painted. Some will show you the various body styles produced by one car company each year. Some will describe the step-by-step method of repair and maintenance of particular cars. Still others will give you a brief overview of a marque through a mass of color pictures.

This book will do none of those things. This book tells the story of a great American classic, the Cadillac Eldorado. Sections of the book describe the cars themselves as they were produced from 1953 through 1994. Other sections of the book contain interviews with designers, engineers, and executives who have worked on these masterpieces.

As the philosophers say, history is like a stream—it flows continually. We have had the opportunity to talk to many of those men who lived at the source of the Eldorado story. Time passes and such an opportunity will never exist again. When we last spoke to the late Bill Mitchell, we did not know we would be doing this book. There are a million questions we *could* have asked him, but that opportunity had passed before we began. We never had the pleasure of meeting Ed Glowacke or Harley Earl before they passed away, but their contributions to the history of the Cadillac Eldorado are monumental. Even though we never spoke with them, we feel we almost know them from all the conversations we've had about them with people who knew them well.

After reading this book, we don't think you will ever look at an Eldorado in the same way. You will also have a greater appreciation for the heritage of today's Eldorado and the work and thought that went into developing these impressive machines. Today, Cadillac is reborn and Eldorado leads the way!

This Book Was Not Written in a Vacuum

This book would not have been possible without the help of many kind and resourceful people. First, we would like to thank all those who took the time to be interviewed by us and review transcripts of texts. Second, we owe much gratitude for the patience and help we received from our editors at Motorbooks International, Michael Dregni and Zack Miller, and their former colleague, Barbara Harold.

Although we bear full responsibility for any errors, special thanks are due the following people, without whose assistance this book could never have been done: Dave Holls, Dick Nesbitt, Franklin Q. Hershey, Floyd Joliet, Stan Parker, Wayne Kady, Vince Muniga, Steve Gaut, Chuck Jordan, Sergio Pininfarina, George Moon, Leonard Cassillo, Dan Adams, George Ryder, Brooks Stevens, John Hambrock, Herb Rothman, Ted Davidson, Carol Jacks, Tara S. Howell, William H. Storm, Richard M. Langworth, Vibe Klarup, Jorges Ramirez, Frank Peiler, Rich Taylor, Greg Wallace, Alan Haas, Michael Lamm, Allan Dowling, Dave Tellepsen, Pierre Ollier, Jim Pfeffer, Duane Medley, and, in particular, Roland and Joan Maki.

Bill and Jeanna Howell
October 14, 1994

American Classics

Chapter 1

Background for a Legend

"The Standard of the World" is not just an empty motto for Cadillac. The motto and the first Dewar Trophy were awarded to Cadillac in 1908 because it was the first automobile to pass a rigorous test of its precision interchangeable parts. The birth of the Eldorado in 1953 was a significant milestone in the history of the marque. This luxurious concept did not appear overnight, but evolved out of the proud succession of Cadillac's engineering and design triumphs. The Eldorado is the standard against which all American luxury cars are judged.

Three earlier Cadillacs set the stage for the emergence of the Eldorado: the 1930 V-16 452, the 1938 Sixty Special, and the 1948–1949 Cadillac.

Three giants in the world of automotive design laid the groundwork for the birth of the Eldorado: Harley Earl, William L. Mitchell, and Franklin Q. Hershey.

Harley Earl established the tradition of design excellence at General Motors when he joined the corporation in 1926 as a consulting engineer. For his first project, he designed the 1927 LaSalle, largely based on the Hispano-Suiza and some ideas from the firm of Hibbard and Darrin (builders of custom automobile bodies). It was such a success with Alfred P. Sloan and the other executive powers at GM that Earl was asked to permanently head up a newly established Art and Colour Section. This section later became the

1931 Cadillac V-16. Unfortunately, the first of these supersmooth-running automobiles was introduced just a couple of months after the Wall Street Crash of 1929. In spite of that, they remained popular with the few people who could afford them. Sergio Pininfarina's father built a special boat-tailed version of this beautiful machine in 1930 at the beginning of his company's existence. The V-16 was Cadillac's first prestige special production automobile using a philosophy later used with the creation of Eldorado.

1933 Cadillac V-16 Aerodynamic Coupe. Considered the industry's first real show car, the flowing lines of this beautiful machine wowed visitors to the General Motors Pavillion at the 1933 Chicago World's Fair. Harley Earl went against the boxy style then in vogue and even hid the trunk and the spare tire inside the sweep of the rear end. The gas filler cap was integrated into the taillight, an innovation that first appeared on production Cadillacs in 1941. The long accent line of the car drew the total form together, making you forget that the front end was basically stock.

William L. "Bill" Mitchell's first masterpiece. The 1938 60-Special was the first production car without running boards, but more importantly, its compact and integrated design set the course for personal luxury in the automotive industry.

GM Design Center, after having been known as GM Styling and GM Design Staff.

Until his retirement in 1959, Earl oversaw the design of all GM products with a firm hand. A towering legend, both physically and by reputation, in GM's styling studios, Earl's eye for design excellence was instilled in all the company's designers, either through intimidation of the stylists who worked for him or through the awe and respect people had for the man.

Earl had been an automotive designer in California where he styled custom bodies on such production chassis as Rolls-Royce, Pierce Arrow, and, of course, Cadillac. Indeed, he worked for Don Lee, Cadillac's high-volume California distributor. Lee had bought Harley's father's Earl Carriage Works and was producing custom bodies for West Coast movie stars and millionaires, with Harley Earl, the junior, as primary designer. Their illustrious customers included the likes of Tom Mix, Mary Pickford, Cecil B. De Mille, and Fatty Arbuckle.

Even though he was *the* power at GM, Alfred P. Sloan was a bit timid about releasing Earl into the midst of the entrenched engineers and other established professionals of the corporation. Even as head of a new section created personally by Sloan and blessed by the powerful Fisher Brothers, Earl had a long road of tough battles ahead to establish his authority at GM. Cadillac-LaSalle's record sales in 1927 and 1928 helped this newcomer's reputation.

Earl became not only the force behind automotive design at GM, but also the founder and developer of the profession of automotive design as we know it. For Earl, automotive design was a process employing the efforts of the best talent he could obtain.

Lawrence P. ("L.P.") Fisher, Jr., the president and general manager of Cadillac Motor Division, had an idea for the production of a "super car," a car that would place Cadillac on a completely different plane from its luxury competition

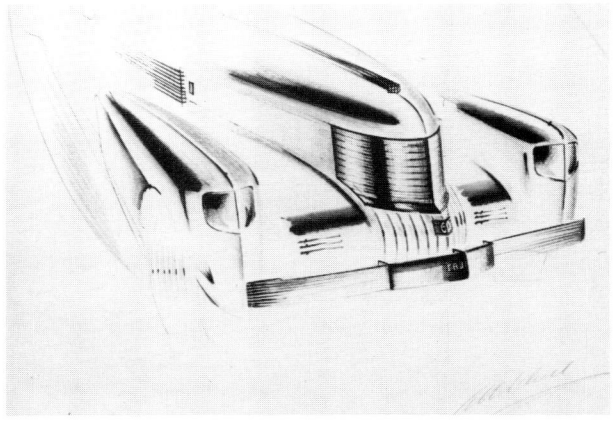

One of Bill Mitchell's early design studies after joining GM. Looking at the rendering, you can see that the components of his work are not just parts of an automobile piled together, but an integrated structural design that works as a unit.

and from its sister automobile, the LaSalle. The incredible V-16 Cadillac, known as the Series 452, was the physical product of his super-car concept. The bad news was that the car was announced shortly after the fatal stock market crash of 1929. The good news was that it was graced by a sleek, low body styled by Harley Earl himself. Earl always said he spent a great deal of time turning squares into rectangles. The low-slung body of the massive V-16 was no exception. Its luxurious appointments, magnificent coachwork, and the specially developed engine make the V-16 one of the high points of the classic era.

The massive engines for the V-16 were designed by Owen M. Nacker who demonstrated his respect for this project by specifying that each bolt and pipe on the engine be chrome plated. Indeed, the appearance of this behemoth powerplant—it and the V-12 were the first engines to receive a styling treatment—was just as carefully planned as the mechanical engineering. Wherever possible, the wiring was hidden. The black aluminum rocker covers were polished and finned, which also resulted in more efficient heat dissipation.

Bill Mitchell with the 1963–64 Buick Riviera he had built for his mother. We took this photo at his home when he lived just outside Detroit.

Mitchell was a driving force in GM design for decades and a major player in the Eldorado story. His design legacy cannot be over emphasized.

During development, security was tight. On purpose, Nacker and his crew were pretty lax about letting it be known that they were perfecting a V-12. In fact, he was doing this at the same time. But the V-16 was a complete surprise to the industry.

Only a few of these super cars were built. The beginning of the Great Depression was just not the right time to introduce a mammoth automobile that returned fewer than 8mpg and at a purchase price which at the time would have supported a large American family for several years.

L.P. Fisher left Cadillac in 1934, the middle of the Great Depression. The V-16 Cadillac flagship stayed in production in one form or another through 1940. By that time, William L. (Bill) Mitchell's influence on Cadillac styling was in ascendance.

At our first meeting with Bill Mitchell, he said he was proud to have "gasoline in his veins." The man loved cars and gave little praise to those in the automotive business who were not real "car people." He began drawing cars at hill climbs and races in the Pocantico Hills and at the Sleepy Hollow Ring held by the Barron Collier family and the Automobile Racing Club of America. In fact, he designed the insignia for this organization, which eventually became the Sports Car Club of America (SCCA). When Harley Earl saw his drawings, he invited Mitchell to join him at GM. Mitchell started work at GM on December 15, 1935.

Mitchell was like a breath of fresh air at Cadillac styling. For Cadillac, the momentum from Earl's arrival had ground to a halt under the pressure of the depression. Earl immediately set Mitchell to work on the 1938 Sixty Special. Bill Mitchell said that Earl told him what he wanted in the car, but made no sketches. Originally the Sixty Special was to be a LaSalle and even bore LaSalle regalia in early design models and sketches, but fortunately it was destined to be born a Cadillac. It was also Mitchell's first masterpiece.

Following Earl's criteria, this elegant sporty machine was styled to look like a convertible and be of a slightly smaller size than other Cadillacs. In fact, Mitchell said the car represented the first of the "youth image." Chrome was at minimum on the Sixty Special; the first full-size production car without running boards. With Mitchell, horizontal lines were overcoming vertical ones.

The importance of this car's design cannot be overstated. Bill Mitchell's Sixty Special set the tone for most of the remainder of Cadillac design history. This was a personal luxury car of a new age; a luxury car that did not necessitate a chauffeur. This was transportation for the successful individual.

Mitchell's influence washed over all the cars in the Cadillac line. He was not only the head of Cadillac design from 1936 until he left for service in World War II, he *was* Cadillac design.

Around the beginning of the war, Harley Earl obtained permission to take some of his designers out to Selfridge Field Army Air Force Base near Detroit to have a look at some of the then super-secret aircraft. One of the planes the designers saw was the Lockheed P-38 Lightning, which would later become the terror of German and Japanese wartime aviators.

The P-38 was a twin-boomed, twin-engined fighter of beautiful proportion and line. The plane's designers, Hall Hibbard and Kelly Johnson, said the plane had almost designed itself once they had been given the performance specifications by the military. Earl's designers were impressed by the aircraft and worked many aspects of its contours into their design studies over the next few years.

Among the designers who went on the excursion were Bill Mitchell and Franklin Q. Hershey. Mitchell recalled that what impressed him about the plane most was the flow of the single unbroken line from the cowl all the way back to the tip of the tail.

Many of the design studies that reflected what had been seen on this field trip were outlandish and uncommercial. But a couple of ideas were not, including the continuity of unbroken line, which Mitchell had admired, and the tailfin. (Frank Hershey eventually brought the tailfin from the P-38 to its place in Cadillac design history.)

Following his stint in the military, Hershey became head of Advanced Design. Because Mitchell was still away in the Navy (and subsequently the head of a private design firm owned by Harley Earl), Hershey became head of Cadillac design for a time after his service in the war.

Some histories of Cadillac make the error of setting Mitchell's tenure as head of the Cadillac studio from 1936 to 1949. But he was either in the Navy or working in other capacities during much of that period. According to GM Styling Section memos confirmed by people who were there, Frank Hershey, Arthur Ross, and Joseph R. Shemansky also occupied that position at different times during the period.

Franklin Q. Hershey had first been hired by Harley Earl in 1928. He went back to Murphy Body in October 1929, and returned in 1933 to stay with GM until 1949. He served in the war as an officer in the Navy, but rejoined GM in 1944. His early work at GM included a stint as head of Pontiac design where he was responsible for the famous "Silver Streak" Pontiacs—a styling motif and identifying characteristic of Pontiacs for about a quarter century. Hershey's designs had lifted Pontiac from near extinction to record sales. For about three years, Hershey was the chief stylist for GM Overseas Operations. This included a stay with GM's Opel Division in Nazi Germany, and projects in Australia and England. After GM, Hershey worked for Packard and then Ford where he was responsible for the 1954 through 1957 Fords. From his time there, he is best remembered for his design of the 1955–1957 "little" Ford Thunderbirds.

Hershey came out of the same world of custom-bodied California cars as Harley Earl. For several years he was the chief designer for Murphy Body, creating custom bodies for Minerva, Rolls-Royce, Duesenberg, Lincoln, Peerless, and Cadillac. When Harley Earl brought him to GM in 1928, he viewed the young Hershey as part of the foundation of the corporation's design program.

After the war, Cadillac, like most automobile manufacturers, had to make do with a superficial overhaul of their 1942 designs and hang a 1946 model year designation on them. During the war, the Cadillac assembly lines produced tanks, and, ironically, engines for the P-38s, among other

Harley Earl (right) with Franklin Q. Hershey at a 1946 GM "Art & Colour" western picnic. Automobile design and the process of automobile design are what they are today because of Earl. He was the man who organized the design function at General Motors and set up a system for producing designs that would last. The Eldorado line was launched on his order for 1953 because he thought Cadillac needed "something special."

things. There was a postwar race in the industry to come up with designs that were completely new, ideas that would satisfy the pent-up demand of a war-weary public who, because of the conflict and the Great Depression, had not had the opportunity to buy cars for about fifteen years!

Having returned from the Navy before Mitchell, Earl told Hershey to produce an all-new Cadillac for 1948. He oversaw two programs, one of which was a dead-end run with a design that the corporate heads said was too far ahead of its time. Hershey also worked on an idea that had come to him when he saw the P-38 for the first time before the war.

After setting the idea in clay, Earl saw it and was not happy.

The Lockheed P-38 Lightning, the plane the Germans called "The Fork-Tailed Devil." Harley Earl took a group of designers for a look at this then-restricted aircraft at an airfield near Detroit. Bill Mitchell was impressed that a single unbroken line could be traced from the tip of the prop rotor to the tip of the tail. Aeronautical designers Hall Hibbard and Kelly Johnson said that once they were given the specifications from the military, the plane pretty much designed itself. Frank Hershey adapted the tail to the tailfins of the 1948 Cadillac. Ironically, Cadillac built engines for the plane during World War II. *Photo courtesy of Lockheed-California Company and Mr. Jeffrey L. Ethell*

Pointing to the tailfins on the clay model, he expressed his displeasure. "That just isn't Cadillac," Earl said to Hershey. "Take those things off!"

Anyone who knew Frank Hershey knows that he was and is the epitome of the rugged individualist. He had great respect for Harley Earl, but he knew good design when he saw it. In this case, he just covered up the rear of the model with canvas and let the fins be. When he believed in an idea, he stuck with it.

Later, Earl came into the studio and saw the model again. It still sported the same distinctive tailfins.

"Frank, I thought I told you to take those things off!"

The next time Earl came into the studio, the managing director of Cadillac, Nicholas Dreystadt, and a couple of important district sales managers accompanied him.

"Frank, I hope you still have those tailfins on the car; everybody loves them," Earl said.

Frank Hershey uncovered the model, which, of course,

Left, Franklin Q. Hershey and the authors' dog, Waggy, in 1990. A massive strike against GM right after World War II led Hershey to set up makeshift basement studio at his farm outside Detroit. He set the design for the 1948 Cadillac during this period. The design theme carried over into the early Eldorados. Hershey left GM in 1949 and was later responsible for the design of the 1955–57 Ford Thunderbirds. Today he is retired in California where he does volunteer work at a nearby railroad museum.

Hershey and "Cadillac Motor Division - GM Styling Section" at his farm outside Detroit while they were working on the 1948 Cadillac design. Hershey is at the left, sitting on the door at the top of the stairs. The man at the bottom with the cigarette in his mouth was the Chief GM Sculptor, Chris Klein, the man responsible for the beautiful hood ornaments and other details of the '48 and many other Cadillacs. Bill Lang is standing opposite Hershey with a cigar in his hand. The other men are sculptors and modelers on the design team.

still sported the tailfins. Later, Earl said that Dreystadt and the sales managers had loved the fins. Thus, an idea that Frank had toyed with for quite a while was finally approaching the approval stage. In fact, Hershey had used the rudder-like tailfins on some of the early drawings Ned Nickles drafted for the Advanced Design Studio.

Shortly after Hershey's confrontation with Earl and Dreystadt, there was a lockout at the General Motors Building and Plants. Frank Hershey took his Cadillac design team out to his farm where they worked on the designs for the 1948 Cadillac in the basement of his farmhouse. The small group even made a sign which said "Cadillac Motor Car Company—GM Styling Section" and hung it over the entry door to the basement and on the farm's outhouse.

What emerged from Hershey's farm was the all-new 1948 Cadillac—the first Cadillac with tailfins. Advertising for the 1948 model referred to them as "rudder-type" fenders, but the public fell in love with this new Caddy with "tailfins."

Actually, to say this was the first Cadillac with tailfins is to oversimplify. Bill Mitchell maintained later that Cadillac had used the tailfin earlier when the chrome and plastic raised taillight had been added to the rear fender, as in the 1941 model. But that was not really the same thing. Hershey's tailfin was an integral part of the rear quarter of the automobile. Its upswept flow of line actually made the front of the car appear lower. From the front three-quarter view, the effect was enhanced by the horizontal lines of the front grille and bumper.

The massive horizontal theme of the front had been a characteristic of Cadillac since 1941, and is a recognizable feature of the marque to this day. On the 1948, the grille perfectly balanced the rear fin. The proportions were exquisite. Cadillac was long, and low, and had the feel of luxury—all the things that Harley Earl had said Cadillac should be.

The flow of line was there too, that long, unbroken line from the front to the rear that Mitchell told us so impressed him when seeing the Lockheed P-38. Even the outward relief bulge of the rear fender does not break the flow and, in fact, faintly recalls the projection of the radiators on the fuselage booms of the airplane that inspired the car.

For 1949, the Cadillac's appearance changed little except for the interior and a few cosmetics (some traditional Cadillac customers had not been comfortable with the 1948 dashboard design). One major exception was the reintroduction of the clean-looking pillarless hardtop. Buick had presented their pillarless hardtop in the Riviera of 1948, but Cadillac had not produced this bodystyle since 1928. We give a nod here to the 1928, but in reality the 1949 was Cadillac's first true realization of the body style.

1948 Cadillac Prototype. Frank Hershey and his group considered several futuristic designs before the final development sequence for the design of the 1948 Cadillac. This is one of many. Some had tail fins and some did not. Note the curved windshield. This particular design study carried the name "Vauxhall," but is quite similar to others in the series that had completely different names. Pillarless hardtops were also represented in this set of design studies. Full-size prototype cars featuring some of these themes were built and actually tested at the GM Proving Grounds. Finally, management decided the designs were too futuristic for public acceptance and a more conservative design course was set toward the bodystyle that actually became the 1948 Cadillac production model.

The big news for 1949 was under the hood. Cadillac was no longer just a pretty face—it now had real muscle.

The 1948 Cadillac was the last in which the old L-head V-8 was used. Before the war, GM engineers had realized they had just about reached the limits for expanding the power of this old design. This was primarily because when the compression ratios were increased, the breathing restriction between the valves and the cylinder bore was also increased. Cadillac had introduced a V-8 in quantity in the 1915 model. The time had come for some substantial changes.

Overhead valves solved the breathing problem. This change and many others were embodied in a totally new "high-compression" V-8 designed under the supervision of three engineers: John F. Gordon, Edward N. Cole, and Harry F. Barr.

The combat needs of World War II had forced engineers to stretch their abilities to come up with materials that would withstand the pressures and speeds necessary for high-performance aircraft. One result was the development of a new aluminum alloy that could lengthen the life of engine bearings by a factor of three. Used in the new V-8, along with other fresh ideas, GM engineers were able to design an engine that possessed a valve and lifter life increased by 50 percent and an overall longer-lived engine.

In its developmental form, the new overhead-valve engine's displacement was reduced from 346 to 331ci. Horsepower increased from 125bhp (brake horsepower) at 3200rpm to 135 at 3500—all in a lighter, more durable, and smaller engine that became the envy of the industry. The engine was such a success that only small modifications and displacement increases were made before the next new engine was unveiled in 1964!

Motor Trend pronounced the 1949 Caddy the first "*Motor Trend* Car of the Year."

This marvel of the engine designer's art quickly became the racing engine of choice among some of the best names on the world's tracks. With hardly a second thought, world-class racers would pop the new Cadillac engine into whatever car they preferred. The result was a series of racing victories for such amalgams as "Fordillacs," "Cad-Allards," and "Studillacs" up through the early fifties. Briggs Cunningham even raced a Cadillac-engined car with a body built by Grumman Aircraft affectionately known as *Le Monstre*. Cadillac power was what you needed if you wanted to win.

If you purchased a new Cadillac during that time, your neighbors knew that you had the latest in engineering. Not only that, it was packaged in the most beautiful luxury body on the market.

If you park a 1948 or 1949 Cadillac in front of a 1953 Cadillac Eldorado, the kinship is readily apparent. There can be no mistake, that elegant golden line is present in both cars.

Cadillac's reputation was set for the introduction of a new line of automobiles that would be the cream of the cream. The public expected even more from Cadillac and from General Motors. A place had been set for the introduction of the Cadillac Eldorado.

The 1949 Cadillac with the all-new V-8 engine. Much like the 1948 Cadillac in appearance, the 1949 was powered by a new engine of such power and efficiency that it was used on race tracks across the country. Compare this picture to the first Eldorado of 1953 and you can immediately see the kinship between the two cars.

American ♦ Classics

Chapter 2

1953: The First Cadillac Eldorado

In 1953, General Motors was riding high. Cadillac was eleventh in overall car sales, it received the 1952 Car of the Year Award from *Motor Trend,* and there were a pile of 1952 model year orders that had to be filled with 1953s. Obviously, Cadillac was becoming the primary luxury choice of the buying public. In 1952, 36 percent of the people who had bought a luxury car had bought a Cadillac.

Nineteen fifty-three started with pomp and circumstance in January when Dwight David Eisenhower was sworn in as the thirty-third president of the United States. In

Left, early concept rendering by Ed Glowacke. Dave Holls remembers this painting hanging in the studio when he joined GM well before the order came down to do the Eldorado. Glowacke's art and attention to detail and drama are both apparent in the rendering. Mitchell said he was responsible for the Dagmar (or "bombs" as they were sometimes called) bumpers. Holls said that Glowacke took the time to show new designers how to design a car.

The Buick XP-300, pre-Motorama show car took four years to design and build. Its wraparound windshield caught Harley Earl's fancy, and he wanted to see it on a production car. The Cadillac Eldorado concept was the thing that could bring that wish into reality.

1953 Cadillac LeMans. With this beautiful Motorama show car, Cadillac was telling the world what was to come in the immediate future. The Eldorado themes were developed and presented behind elegant, hooded headlamps. If you were visiting Motorama when you saw this car, it was reassuring to know you could go to your Cadillac dealer and buy the 1953 Eldorado which had many of the same design characteristics.

the inaugural parade, he and his wife rode in the new 1953 Cadillac Eldorado. This was the first time most Americans had seen or heard of this phenomenal new Cadillac.

GM Motorama

In addition to presidential appearances, there was something called the General Motors Motorama touring the country, stirring up enthusiasm for automobile dreams yet to come. When Motorama came to our town we went at least once a day, every day it was there. Motorama's GM dream cars, showing what was possible and what might be in the future, drew crowds across the nation. The cars of Motorama woke up the American public. Suddenly the fine automobiles of just a few years ago didn't seem so exciting. A curtain had been pulled aside to yield a glimpse of a bright future.

Actually, the future of Cadillac design was hinted at in pre-Motorama days by two beautiful convertible show cars, the Buick XP-300 and the LeSabre, unveiled at the Paris Auto Show. Both of these rolling automotive laboratories took about four years to design and build. These were not just pretty skins pasted over production chassis. Buick's chief engineer, Charles Chayne, planned for the XP-300 to burn two fuels, gasoline at low speeds and methanol at high speeds. Both cars had built-in hydraulic jacks and myriad special features.

1953 Cadillac Orleans, a show car. Called the first American automobile with four-door pillarless construction (the rear doors were to open from the front), this car carried a massive amount of chrome on its leading surfaces.

1953-plus Cadillac Le Mans, a re-worked show car. The 1953 version of this car had 1954 Cadillac production-style tailfins and was a part of the 1953 GM Motorama. Actually, much of the design of the 1953 version of this show car appeared in production, especially on the Eldorado. In the later version pictured here, the tailfin essentially incorporated the entire rear fender. The unchromed dual air scoops make this a much more subtle design. This version was only shown a limited number of times and is said to still exist at GM.

We remember seeing the LeSabre on television in the early fifties parked beside the field during a football game. When the weather produced a sudden sprinkle, a moisture-sensitive panel between the seats activated the roof mechanism and the top rose automatically.

The LeSabre was ostensibly a Buick show car—its sweep of chrome from the front wheelwell downward clearly said "Buick"—but its tailfins and Dagmar front bumpers said "Cadillac."

Both cars had Harley Earl's curved windshield. GM designer Dave Holls remembers that Earl was quite proud of this windshield and wanted to get it into production as soon as possible. Holls says that it was Earl's desire to get the windshield into production that led to the introduction of the Eldorado. Earl went to the head of Cadillac, Don E. Ahrens, and told him that his division needed a super convertible and that this new windshield would make the car distinctive.

The Motorama Cadillac show cars, which had the panoramic windshields, revealed much to the public about the trends in the marque's design. The Orleans was much like the 1953 Eldorado and the two-door Coupe DeVille, except that it was a four-door hardtop with suicide rear doors (doors that opened from the center post).

The LeMans was more radical and provided more of a futuristic look. The squared-off rear end and face of the front end was a preview of the 1954 Cadillac. The abbreviated vertical air scoop immediately before the rear wheels and the finned turbine wheels hinted at Caddies that would arrive in a few years. The overall tight design prefigured what Eldorado would be in the mid to late fifties. The finned turbine wheels were used by Cadillac in numerous variations for years, and Pontiac picked up the LeMans name for one of its smaller models in the sixties.

To say that these two show cars "inspired" the first Eldorado or subsequent Eldorados does not accurately reflect how the automotive design process worked in those days. Generally, it took from three to four years from the initiation of the design process until the new car finally rolled off

Ed Glowacke, Chief Designer, Cadillac Studio from 1951 through 1957. Harley Earl gave the order for a special Cadillac, and Glowacke carried out the order and produced the Eldorado. Extremely creative, well liked, admired, and talented, Glowacke was a true Renaissance man. He loved cars and adventure. He pioneered many sports, such as skydiving, in the fifties. He never really left Cadillac Studio because he became Bill Mitchell's assistant under Mitchell's tenure as vice president for Design. Glowacke knew he was in failing health and lived his life to the fullest. Tragically, he succumbed to leukemia at an early age in 1961.

the assembly line. Simply speaking, this meant that the Eldorado was developed concurrently with the two 1953 Motorama show cars rather than being inspired by them. (Actually, there is evidence that the 1953 Eldorado itself was preceded by a 1952 Motorama special show car convertible (having no name) with a wraparound windshield and a straight beltline.)

In fact, development of the car's design was actually more evolutionary than one would think. Since the XP-300 and the LeSabre were in design stages for about four years before their debut in October 1950 at the Paris Auto Show, their panoramic windshields, lowered bodies, chopped doors, and flat rear decks had been around for quite some time. Ed Glowacke, who was head of Cadillac design from 1951 to 1957, developed the distinctive Dagmar bumpers (incidentally, these were named after a well-endowed blonde actress, a regular on the "Milton Berle Show") over a period of time from unobtrusive bumperettes of the 1940s Cadillacs to the huge projections that they became in the mid-fifties. Their presence on the 1953 Eldorado was just a step along the way. Harley Earl was not enamored of them because he felt they distracted from the "gull wing" form of the upper portion of the horizontal bumper.

Many of the show car innovations were more than just dreams. Amazingly, if you had enough money, you could go into a Cadillac show room and buy a car that had a feast of Motorama features. You could even buy a copy of the car the president had just ridden in at the inauguration! That low-slung beauty could be *yours*. The only trouble was the $7,750 price tag—more than the cost of any other GM postwar car. In 1953, you could buy a house for that much money! Needless to say, only a small number, 532, were sold.

An Eldorado in the Showroom

Although the Eldorado did not sell in large numbers, it did sell other Cadillacs. People went into Cadillac showrooms to lust and leer over the new Eldorado, and many drove out with a lesser priced model. It was truly the "golden one." (The name Eldorado had been suggested by one of GM's secretaries and was a perfect fit for Cadillac's flagship.)

The Eldorado did just what Harley Earl wanted it to do: It extended Cadillac's bodystyle program for a fourth year and introduced the wraparound windshield.

If you bought the Eldorado, what you got for your money was a chopped and channeled custom car almost 300 pounds heavier than the standard convertible. No Eldorado script distracted from the exterior and the only insignia was a gold "V" under the Cadillac crest, one on the hood and one on the trunk lid. The Eldorado insignia was reserved for the center of the dashboard and the script for the doorsills. The interior was pleated leather.

1953 Cadillac Eldorado Brochure.

1951 Buick LeSabre show car. In this rare photo of the LeSabre top mechanism in action, one can clearly see the cut-down door treatment, Dagmar bumpers, and curved windshield later seen on the 1953 Cadillac Eldorado production car.

Harley Earl's 1938 Buick Y-Job. Built on a 1938 Roadmaster chassis and designed mostly by George Snyder, this first GM "dream car" developed because of a close friendship between Earl and the head of Buick, Harlow Curtice. When the top was down, the rear deck was flat, forecasting Eldorados to come. Hidden door hinges and headlights accented by a wide front grille were just a few of the car's innovations. Earl wanted the car to test the reaction of the public. That reaction was positive, and he used the car for his personal transportation from 1942 to 1944.

Like Harley Earl's 1938 "Y-Job" and both the 1951 XP-300 and LeSabre show cars, the Eldorado lacked a hump behind the rear seat to conceal the lowered top. When the top was down, a flush smooth metal cover, the same color as the car's body, concealed the convertible mechanism. This was Harley Earl's disappearing top raved about in the brochures.

In fact, the hidden top and the Eldorado's cut-down beltline were the two most obvious exterior distinctions between it and the vanilla Cadillac Sixty-Two convertible. In fact, the Eldorado was considered a part of the Sixty-Two model line until the introduction of the 1957–1958 Eldorado Brougham and the 1959 model year.

Along with the sister special-issue Oldsmobile Fiesta and Buick Skylark, the Eldorado was the first production car with the panoramic or wraparound windshield introduced earlier on the 1951 LeSabre show car and two 1953 Motorama cars: the Orleans and the LeMans. The wraparound windshield both necessitated and enhanced the appearance of the distinctive Eldorado windwings. The rest of the Cadillac line made do with the old windshield.

Standard equipment wire wheels further set off the Eldorado. You could even justify part of the cost of this super car by telling yourself you *needed* the wire wheels to keep the brakes cool.

As mentioned, GM was making use of, for one last time, the dies for this model run that had begun in 1950. Though based on the Sixty-Two, the Eldorado's body panels just didn't match up with the Series 62 line and thus every last one of the 532 copies had to be practically hand built. The dropped-door design gave the car a distinctive lowness, three to four inches lower than the Sixty-Two Cadillac convertible. Taken together, these custom design features made possible the unique beltline dip at the rear of the door.

Special Eldorado features were many. Other Cadillacs did not have the Eldorado's plastic royalite anti-glare dash cover, color-matched with the car's interior. The Eldorado steering wheel also had plastic-leather handgrips that matched the car's pleated leather upholstery. Other standard equipment included whitewalls, built-in fog lights, spot lights, license frame, power steering, a windshield washer, vanity and side-view mirrors, and a signal-seeking, pre-selector radio. Cadillac's Eldorado brochure also listed such things as a heater and an oil filter as standard equipment, things that today we just assume will be on any car we buy.

The American Boulevard Sports Car

Cadillac presented this package as an automobile that "meets the full needs of an American sports car." Here GM was putting the emphasis on appearance, because this was also the year Chevrolet Motor Division introduced the Corvette, GM's answer to the exotic European marques—such as the Jaguar XK-120 and the MG—then arriving on our shores. In producing the Eldorado, Cadillac was presenting an overblown, oversized sporting car for the well-heeled casual American who demanded to retain as much comfort in the bargain as possible.

Comfort was here in abundance. A full sixty-three inches of hiproom in the front seat meant any driver and front seat passenger wouldn't feel cramped. The recesses in the backs of the front seats allowed for extensive leg room for whoever sat in the back seat. The rear seat passenger had fifty-one inches of hiproom and luxurious armrests. One could "sport" about in comfort.

While the rubberized-cloth Orlon top came only in black or white, the brochure said the buyer had a choice of any of the twelve standard Cadillac exterior colors—or one of the four special colors made available exclusively for Eldorado. From all accounts, no one chose any of the standard twelve colors. For interior color choices, the buyer could choose one of three solid tones of leather or one of three combinations of two colors.

The 1953 Eldorado was an outrageous American boulevard sporting car, large and showy. The public loved it. The Eldorado was ostentatious, but, in a way, slightly restrained. Actually, looking at the 1953 Eldorado today, can remind one of the austerity of the 1993 model. Except for the barrage of chrome on the front bumper, plating was minimal (by 1953 standards), and there was little exterior signage blaring out the name "Cadillac." The same is true for the 1993 version. Both cars are quiet about their name, making the assumption that the design of the car says it all. There is no mistake that these two cars–forty years apart–are both distinctively Cadillacs.

1953 Cadillac Eldorado. The first Cadillac Eldorado was available only as a convertible with chrome wire wheels standard and the most powerful engine in the industry, 210hp. Although each had to be practically handmade by Cadillac, these were production cars, and more than 500 were manufactured selling at $7,750 a copy. Aimed at a younger buyer than the traditional Cadillac customer, the car introduced the wraparound windshield and had unique sliding handle door openings.

1953 Cadillac Eldorado. This example of the first Eldorado was purchased in rough shape (totally burned) by its owner, Bob Knapp. Knapp employed Brad Dunn to oversee the restoration of the automobile to its present state of grace and elegance. The car was number sixty-four out of a total of 532 built. Discovering that Eldorado seats differed from those of other Cadillacs, special squared-off backs for the front seats were built to factory specifications to give the car its original sporty appearance. This Eldorado is on permanent exhibition at the Deer Park Auto Museum in Escondido, California.

Dave Holls Interview
The First Eldorado and More

Dave Holls joined GM in 1952 after graduating from Michigan State University with a major in industrial design. Holls had grown up in Detroit and had won a state award in the Fisher Body Craftsman's Guild competition. He worked as a designer in several GM studios before occupying various executive designer positions, and contributed to the design of several Eldorados during his career. His continual interaction with Cadillac designers constituted a major contribution to the design of the current Eldorado and Seville. Holls recently retired from GM.

Author: How did the first Eldorado program come about for '53?
DH: It was Mr. Earl. He kind of set up a competition between Buick, Oldsmobile, and Cadillac to do special cars. You know, Buick did the Skylark and Oldsmobile was sort of forced to do the Fiesta. We didn't—honestly—go in the other studios. They were off limits. Harley just wanted something special in Cadillacs. We hadn't done anything special in a long time as far as he was concerned. We hadn't done anything special since the '48 Sixty Special. That was a different body, you know. We were doing all these Motorama show cars too, but you couldn't buy them. He wanted something that you could buy. The wraparound windshield was it. He wanted to make it available, but it was a terribly, terribly expensive way to go. If this had been something like a new rear end—like on the '55—you'd get a lot more bang for your buck. But you're getting into architecture when you're doing that door. The curved windshield affected the cowl and the doors. It's just an expensive area.

Author: Now, that's because the cowl's the most expensive part of the car to change?
DH: Yes, it affects all other parts—like the doors and that kind of thing. But I'm not sure how much of the '53 Eldorado cowl was affected by the curve of that glass. I just know it was an expensive way to go.

Author: I guess part of that showed up in the price? Did you own an Eldorado when it first came out?
DH: No! No way, I couldn't afford it. I was just a young designer. Yes, I remember Harley Earl came back from buying one. He had bought it for someone else or something. Harley said, "You know, it's terrible, I don't know if you kids will ever be able to afford an Eldorado. I just got one [at a GM discount] for a little over five thousand [in 1953 dollars]. That's crazy!" He hadn't seen domestic prices like that since the V-16 in the '30s. You know, the '53s were pretty much handmade.

Author: The '54 Eldorado was a lot cheaper. What can you tell me about the '54?
DH: The '54 was nothing. The '54 Eldorado was just a piece of aluminum on the rear fenders—you know, compared to the rest of the Cadillac line. That's the way I look at a '54. Some of the collectors don't look at it that way, but it's nothing special. That's a Sixty-Two convertible with a little garbage on it.

Author: And, cheaper than the '53 Eldorado?
DH: But the Sixty-Two was even cheaper and nicer. Now, I wouldn't say that about the '55. The '55 had a very stunning rear end on it.

Author: Who was on the team that worked with you on the '53?
DH: Ed Glowacke was the primary designer. He was the boss of Cadillac Studio and he was a terrific guy. I was kind of a young kid and I kind of worked on it with him. It wasn't a big deal. He was really teaching me how to put a car together then.

David R. Holls.

Author: As I remember, Glowacke died young—in about '62 as I recall. Bill Mitchell always glowed when he talked about Ed Glowacke. He thought he was great. Didn't he think that Glowacke would have taken over after him had he lived?
DH: Yes, without question. Glowacke was the force behind the first Eldorado, no question. The creation of the Eldorado was a technical problem more than anything else, because there was no great redesign done to the car. It was just a matter of getting that wraparound windshield on a '53 production car. To do that you had to change the door—and if you do that, [you might] just as well make some other changes like lowering it a little bit and things like that.

Author: Did you stay with Cadillac through the fifties?
DH: I went in the Army in August of '53 and came back in August of '55. Now we had done the '54 and the '55 before I went into the Army. And, I had done a lot of the '55. I hold a patent for that front side and rear. They actually gave me a patent on it, for some reason.

Author: What was that called?
DH: Well, the side treatment on the '55 begins where that rear vertical vent goes halfway down the fender, then goes forward. Do you know what I'm talking about?

Author: Yes, that's the first difference you notice between the '54 and '55.
DH: That started as part of a sketch of a little sporty Cadillac. When the idea got near to production design, it was originally going to be on the Fleetwood only. We had the cars out at the proving grounds, the Fleetwood and the other Cadillacs. Mr. Earl was there and he said, "You can't tell the difference between a '55 and a '54. You've gotta get that same side treatment onto all of the cars—forget just having it on the Fleetwood." So, Mr. Earl was responsible for having it on all the cars.

Author: So, before you went in the Army, the '55 was done?
DH: Right, I didn't do anything on the '56.

Author: I talked to Ron Hill about his work on the '57.
DH: Ron did the Eldorado for '57, that rear end. I did work on the '58, there was still work to be done on that when I got back.

Author: What was the interaction between the Cadillac designers and what was done in Italy with the Broughams?
DH: Oh, that was a fine arrangement. Mr. Earl had a wonderful relationship with the old man, Pininfarina. And so it was a very nice relationship. It wasn't a competition, let's put it that way. George Ryder was the engineer who went over to Italy on that project. The Italians were given complete lines, but I don't know if models were sent over or not. Now that wasn't a regular Eldorado, that was a Brougham, really. George Ryder could tell you some stories about that project. One story George told me was that the cars were so long, they couldn't get them in their paint booths! So they had to paint the rear of the cars outside the booth—or they backed them in, I can't remember the whole story. He has a few fun stories to tell about that project.

Author: Did you have any Eldorados during the fifties as personal cars?
DH: Gosh, no!

Author: That's what all the designers I've talked to said. What's the story?
DH: Man, I couldn't afford one.

Engineer George Ryder worked on numerous projects during his career at General Motors. He acted as liaison in Italy between GM and Pininfarina during part of the Eldorado Brougham project. At one time he was the executive assistant to the vice president, International. After leaving GM, he became managing director of Pininfarina North America.

Author: Ron Hill has said that he drove European sports cars during that time.
DH: Yes, I remember he had a Porsche.

Author: He seemed to say that the younger designers felt the old guard at GM was not of their generation and . . .
DH: Well, not Ed Glowacke, though. Glowacke owned a Beechcraft Bonanza and a German glider. And he loved to race cars. He had an Italian Siata sports car that he raced. And when the V-8 Corvettes came out in '55, he raced his. He was not like the rest of them, I'll tell ya. Now, that's what I drove: I had a V-8 '55, that's the old body and the new engine. And when they destroyed the Motorama show cars—the four Corvettes in '54, the Nomad, the little coupe, the fastback, all those—I got the detachable top off that thing, snuck it away. So, I had that top on my production '55 Corvette. It was beautiful! Because it was so low, there was little headroom and I had to reduce the seats' height by tying them down. The Corvette's top was so low, you had to pull the seats right down.

Author: The special nature of the Cadillac Eldorado went out the window with the '59s. Was there a reason for this?
DH: It wasn't from lack of interest by Design Staff. There were lots of design proposals for Eldorado. They [GM] just had such an elaborate program as it was to get all those new cars into production each year that it was difficult to do. GM had a new body for every single '59 car, from Chevrolet through Cadillac! With the '59 too, we started a little late. You know the story of seeing the new Chrysler products with the big fins and all, and how that inspired us to do the '59 fins. We had to do a whole new rear on it.

Author: So, do you think that Eldorado as a flagship model was just pushed aside until the 1967 front-wheel-drive model?
DH: No, not at all. Well, Cadillac might look at it that way. Design Staff didn't look at it that way. There were lots of proposals for a new Eldorado, but they just didn't want to spend the money then. Cadillac hadn't made much money on Eldorado—except for the '54. Cadillac probably made a lot of money on the '54.

Author: Were you involved in the front-wheel-drive car for '67?
DH: No. I do know it was brought up from an experimental studio.

Author: What do you consider distinctive Eldorados?
DH: The special Eldorado that comes to mind is the '57 and '58 Eldorado Broughams with the stainless steel roof. That was a very, very dramatic car as far as we were concerned in styling—it caught the imagination of everyone. It cost a fortune to make and GM lost a fortune on every one. It was the first luxury compact, if you think about it. We had that out on the patio with a Mark II Continental. I've got pictures of that. It was striking how modern the Eldorado looked and how old the other looked. Now, don't get me wrong, the Continental had wonderful staying power. But at the time, it looked so old to us. One car on that patio looked like it was from Mars, and one looked like the Duncan Phyfe factory made it.

Author: What is your opinion of the current Eldorado? I know I especially like the interior.
DH: Well, the interior I can at least take partial credit for. You know, it's basically the same interior in the Seville and the Eldorado. People ask me how we got that beautiful interior in those cars. I say, that's simple. I just transferred a Pontiac interior designer into the Cadillac Studio. There was just no way one of the guys in Cadillac who had been there awhile was going to come up with something like we wanted. Marv Fisher, that's his name, I brought him in from Pontiac and eventually he became head of Cadillac Interior. He did a fantastic job!

Author: That interior is almost the definition of elegance.
DH: You know who saw that interior? VonKuhnheim of BMW. He saw it and loved it. Bob Lutz was showing him around the Tech Center. Lutz had been head of sales at Opel when I was in charge of design there. Later, Lutz had been persuaded to go to BMW as director of sales and marketing—a big step up for him. That's where he met VonKuhnheim. Anyway, Lutz talked my boss, Cunningham, into letting Von Kuhnheim come around and look at a real design department, namely the one we had at Opel. Everyone became such good friends and it worked out so well that GM allowed VonKuhnheim a look at design in the Tech Center here in this country. You know, we hid some stuff and things like that, but we did show him around. When we showed him the interior mockup for the Eldorado and Seville, we just couldn't get him out of there. He just sat there and sat there. It is just the exact opposite of a BMW interior, which is about as busy as things get. The new Eldorado and Seville interiors just have everything there that you need.

Author: And the exterior of these two current cars, the Eldorado and Seville?
DH: Did you know that Pininfarina did an Eldorado for us too? I think Cadillac ordered it.

Author: No, I didn't know that.
DH: That was a little different. Cadillac wanted it done. Before that car got over here, it was like a Lutheran in Rome.

Author: Is there something you want to add about the Eldorado story?
DH: One thing is that not too many talk about the '58 Eldorado. You know, the '57 was not very well accepted at the time. They thought it was fat, like a bathtub, back there. The fins are actually too small for the rest of the car.

Author: And you worked on the '58 Eldorado?
DH: Yes. Anyway, Bill Mitchell said let's get some sharpness in that car. It's all rounded; it's all fat. The front is severe and the back's round. So that's why all that sharpness got into it, even the bumper. The individual bumpers go to a sharp corner before they go down the side.

Author: Well, there was some sharpness in the Eldorado, but overall, '58 wasn't a banner year for GM Design.
DH: Oh, let me tell you. Ed Glowacke and I were invited to come down early to the ad agency and look at the roughs they had for the Cadillac print ads. Ed took one look at them and said, "Man, we've gotta get rid of some of this stuff!" We could see it even then. The old man [Harley Earl] was having us put buckets of chrome on those cars! Nobody wanted that crap! We literally had to put the chrome on by the bucket. None of the divisions wanted that stuff—nobody in Buick, nobody in Olds, and nobody in Cadillac! But, that was old man Earl, and he was losing it there at the end. That's why we did the '59s the way we did. We just went too far the other way with the body sculpture.

Author: Which were the most influential show cars for those '50s Eldorados?
DH: First, the LeMans, because from it came the '54 front end, absolutely intact. You'd have to say that was influential. And, the Park Avenue—I think that is what it was called—was just the way the Brougham came out. It was dark green with the stainless steel roof. The '54 El Camino [Motorama show car] was my car. The La Espada came out while I was in the Army. It was the convertible version of the El Camino. I didn't even know they were going to make a convertible out of it. From that front end came the '59 hood. All the hoods of '58 were an old-fashioned hood shape. The El Camino had a pointed hood, so it led the way on the '59 GM cars. The El Camino rear fenders became those of the '55 Eldorado, and taillights were used too. Did you know the El Camino and the La Espada were supposed to be half Ferrari and half Cadillac? The front slanted back underneath and a had thin egg-crate grille like a Ferrari—and not like the heavier egg-crate grilles we put in the '54s. So the El Camino was sort of a little Ferrari/coupe/Cadillac. Of course, it didn't end up that way, it was much more Cadillac toward the end, but it started out that way.

Author: Were you involved in Motorama?
DH: Not really, except for the El Camino design. I was in the Army through a couple of years after that. But I do remember Harley Earl coming back from the '53 Motorama. You know, it had the LeMans in it [the origin of the

'54 Cadillac front end]. Harley said, "You know who wants that car when we get rid of it? John Wayne. He just loved it. One of his friends said, 'Gee, John, you don't want that thing. It's made of fiberglass.' And John Wayne told him, 'Look, I don't care if it's made of puddin', I still want it.'"

Author: So they just broke up all those cars?
DH: No, not the '53s. The '53s were all running cars and they did not break them up. That's the reason that from '54 on, they broke everything up because those '53s got out and they were a disaster! The bodies weren't engineered. Little pieces broke and nobody could replace them—and special paint had to be used. Oh, it was just a nightmare! You know, they sold them just like new cars, and the buyers just assumed they were completely engineered like our production cars. These cars had been built just for show. Just to show what could be done. If a collector today got hold of one of those cars he would expect to put up with all those things that went wrong. He'd understand.

Author: The '54 El Camino was not a running car?
DH: No. Very few of the '54 and '55 Motorama cars were running cars. You know, they had engines in them and all that, but nobody bothered making them run. You didn't want to because the bodies were so bad. There was nothing there to do with structure, it was just enough to make the fiberglass stay together and hold the interior and all of that. They weren't roadable cars. Now, they certainly could have been if you'd wanted to put them into production, but they weren't as they stood in Motorama. That was all the more reason why they didn't let 'em be sold. But it was a sin when they chopped those things up. It was awful. It was an atrocity.

Author: What was the cutoff point between the cars designed at the old building in downtown Detroit and the new GM Tech Center in Warren?
DH: The end of the work on the '57s was at the old building. And I think we did some Eldorado work still out here at the Tech Center on the '57. But it was the '58 that was first done completely at the new building. In fact, the first pictures we ever did on the patio at the new Tech Center in Warren were of the new '57 Eldorado Brougham photographed with the Continental Mark II.

Author: So that was of the '57?
DH: Yes, but they were done. We didn't want to do those pictures at the old building if we could help it. That was terrible! At the old building you had to take the cars up on the roof with all those air vents and ducts. It looked like something out of a tenement in Brooklyn, New York. But the '58 was all done in the new facility. Of course, there's always a little last-minute stuff on any model that has to be done—as I'm sure there was on the '57. I remember that distinctly. We showed the '56s on Dedication Day [May 16, 1956] at the Grand Opening of the Tech Center for the public when we moved in—a couple of weeks ahead of time.

A M E R I C A N ♦ C L A S S I C S

Chapter 3

1954–1960: Fins and Elegance

Motorama Continues to Inspire

Show cars, Motorama, America's first postwar sports cars—all these things were coming together at once to make the people of the United States in 1954 more excited than ever about what was available on the auto market. The Corvette had been introduced the year before. The Thunderbird was announced late in the year, and millions of people had visited General Motors' spectacular Motorama. The promise of what the American car could be was coming true.

The show cars of Motorama were a concrete expression of some of the thinking that was going on in Detroit. Motorama was also Harley Earl's new way of getting his hand on America's automotive pulse. The automobile was the product of a glamour industry that fed Americans' need for freedom, mobility, speed, privacy, and even a bit of self-expression. People were interested in what they could look forward to from Detroit. To own a sporty and personal automobile was no longer a fanciful dream, but an attainable reality.

Harley Earl presented awe-inspiring show cars at the Motorama, watched the public's reaction, and took what he learned back to the GM studios to see what he could build that the public would buy. He systematically mapped the progress of what the Eldorado would be, and began meetings on developing a *super* Eldorado—the Brougham—as early as May 1954.

The 1953 Cadillac Orleans show car was, at first glance, not too different from the production Series 62 Cadillac. But a second look revealed the fact that this ap-

1954 El Camino. Designed by Dave Holls, this Motorama coupe was developed by other GM designers into the LaEspada convertible after Holls left for the service. In altered form, the brushed aluminum roof, aircraft styling, and supersonic tailfins would soon be seen on the production Cadillac Eldorado. The thin horizontal fluting on the front fenders was for venting hot air out of the engine compartment.

1954 Cadillac LaEspada. The striking convertible version of the El Camino drew lots of praise from visitors to Motorama. This topless version emphasized the extreme rake of the windshield and the balance in length between the front and rear of the car. The car appeared in "Apollo Gold" with whitewall tires and painted white with blackwall tires. The balance of design in both cars approached perfection.

1954 Cadillac Park Avenue. This *faux* four-door luxury show car probably had more direct effect on the fifties Eldorados than any of GM's other design exercises. Some of the Cadillac themes found in other show cars were muted in this rendition, while others were exaggerated. The overall effect was one of subdued elegance with a touch of the sporty.

The production 1954 Eldorado Convertible with the top up. Essentially, the 1954 Eldorado took advantage of the beautiful styling of the all-new Series 62 1954 Cadillac. Costing substantially less than the 1953 Eldorado, the 1954 version carried the special aluminum treatment on the lower rear quarter panel, vertical chrome bar lines over the rear bumper, chrome wheels, several badges (including one in 19-carat gold on the dash), and a special leather interior to mark it as an Eldorado. This particular car has had with a continental kit added and was restored and is owned by Herb Rothman and Ted Davidson of Santa Ana, California.

peared to be a four-door pillarless sedan! With suicide rear doors, no less! It was the first showing of this general body type in the United States. Just five years before, Cadillac had wowed the public with the introduction of their two-door pillarless hardtop. Now this. And the roof was different, presenting a more flowing line than the production Cadillac.

The "sports prototype" 1953 Cadillac LeMans two-seater show car was even more striking. Compared to the Orleans, its fiberglass lines were squared off. Its body had a leaner, less bulbous appearance. The restrained and abbreviated scoop on the side at the leading portion of the rear fender was reminiscent of the 1948 and 1949 Sixty Special scoop. Its wraparound windshield and relatively short (115 inch)

1954 Cadillac Eldorado convertible with the top down. An exclusive hard cover, or parade boot, carried in a specially fitted bag inside the trunk when the top is up covers the top mechanism when the top is down. The flow of line from the hooded headlights emphasize this car's long horizontal appearance.

1954 Eldorado Convertible rear seat showing the three-piece body-colored parade boot in place. Owners of the Series 62 convertible had to make do with a soft vinyl snap-in-place convertible top cover. The hard, painted, three-piece Eldorado parade boot looks like part of the car body when in place.

1954 Eldorado rear end. This car is more typical of the model year in that it does not have a continental kit. When Herb Rothman and Ted Davidson purchased this car it was literally a basket case, in that most of the parts were in boxes. Now beautifully restored, it is registered and driven regularly.

1955 Eldorado clay model. This is a fairly late-stage (November 1953) full-size developmental design model clearly showing the astro fins similar to those on Dave Holl's El Camino show car. The break in the rear-quarter panel vertical chrome strip is clearly visible. Designers used a foil material, carefully smoothed, for chrome components such as bumpers and Dinoc plastic sheets, similarly laid down, for painted areas.

1955 Eldorado clay model. An attempt was made to stay with the 1954 rear quarter treatment. Actually, this and the following photographs were made of the same clay model. On the driver's side, as shown in this photograph, the full vertical *faux* vent was tried at the leading edge of the rear fender. Note also that the accent line above the rear wheel sweeps back to actually rest on the top of the exaggerated rear bumper, and the oversize taillight lenses are in the trailing edge of the tailfins. This section was greatly toned down for the production model.

wheelbase emphasized its sporty nature. Four of these cars were built, and its front end and rear squared-off fins appeared on the 1954 line of Cadillacs.

For 1954, Dave Holls created the Cadillac El Camino and, following the custom of some of GM's best designers, incorporated aircraft styling in a beautiful, windswept, elegant package that terminated in sleek, swept-back tailfins. The brushed aluminum roof was a forerunner of the stainless steel roof that would crown some Eldorados in the near future. Horizontal vents in metallic trim panels behind the front wheels allowed the escape of hot air from under the hood. The exhaust pipes incorporated in the rear bumpers became a rocket-line ridge on the side of this 115 inch wheelbase car, extending to a point in the door. Dual headlights accented Holls' pronounced horizontal treatment of Ed Glowacke's more traditional Cadillac front end.

When Holls went into the service, Glowacke and some other designers took the El Camino lines and produced the convertible version, the Cadillac La Espada. Like the El Camino, this car had a windshield that swept back a rakish 60 degrees from the hood. These cars set an idealized standard of what the Eldorado could become; they were possibilities, but three-dimensional possibilities you could examine from all sides.

Also in 1954, Cadillac presented the faux-four-door-hardtop Park Avenue, a luxury show car that would have direct and immediate influence on the Eldorado and Eldorado Broughams of that decade. The massive, but not sweeping, tailfins, the large front-wheel openings, and the long accent line originating with the hooded headlights and dipping into the rear-door vent line gave a stately balance to this 133 inch wheelbase, 230.1 inch long automobile.

All this was part of the climate into which the 1954 Eldorado was introduced. It was a magnificent car, but some see it as a stopgap measure produced to mark time between the almost handmade Eldorados of 1953 and the completely restyled 1955 models that were to come.

1955 Eldorado clay model. The front grille of this model has a heftier treatment than was finally chosen for the production model. The production model's grille more closely fitted the delicate lines in the wheels, the vent under the windshield, and the lines of the *faux* vent at the beginning of the rear fender. With the exception of the lack of a horizontal chrome accent strip leading to the taillights, lack of accent on the top of the door, and different wheels, this was pretty close to the production model.

1955 Eldorado clay model showing an attempt to use a break in the vertical line. This is more like the final 1955 production model in that the vertical line marking the *faux* air scoop at the leading edge of the rear fender is broken halfway down and suddenly goes forward as a horizontal chrome line to the front of the car. Dave Holls developed this modification in design and holds a patent on it. Also notice that the exaggeration of the rear flowing body mass over the rear bumper is more visible from this angle and that the bumper's mass appears to rise out of the back of the rear wheel well in a solid form marked by some vertical lines. This is similar to the rear bumper treatment used in 1957 and 1958 Eldorados. In fact, the vertical lines were picked up for the 1956 Cadillac Eldorado.

1955 Eldorado hardtop fiberglass model. This is close to the production form of the car, but a hardtop on the body shape and chromed wire wheels were tried to see how they fit the design. Notice the 1955 goddess on the hood. In 1956, the first hardtop Cadillac Eldorado Seville was introduced into production, but with a double vertical wing hood ornament.

Practical Matters

Many like to think that the 1954 Eldorado merely takes advantage of the magnificent restyling Cadillac had experienced for that year. Certainly, if you consider the concomitant reduction in the price of the automobile, this seems reasonable. The 1953 was, in some ways, the result of a slow fattening of the basic 1948 design. The 1954 smoothed out those bulging lines and squared off the tailfins, resulting in a more elegant look. The price was easy to swallow too, being $2,000–$3,000 less than the 1953 model.

Differences between the Eldorado and the Series 62 models were few. For $1,300 more than the Series 62 convertible, you were getting a parade boot for the lowered convertible top, a ribbed bright aluminum panel along the bottom of the rear fenders under a gold Cadillac crest, real chrome wire wheels, and a special leather interior. All this, and a few badges (the one on the dash was eighteen karat

1955 production Cadillac Eldorado. The treatment at the top of the door is similar to that found on the 1954. The lines in the sabre-spoke wheels complement the vertical bars on the rear of the car.

1957 Eldorado Biarritz clay model. This photo was taken late in the development of the car's design; most of the car's features have been worked out.

1956 Eldorado Biarritz and Seville production cars. This was the first year for the Seville hardtop and 3,900 copies were sold compared to just 2,150 copies of the Biarritz convertible. Suddenly, Cadillac could see that there was something to be said about sporty, prestigious hardtops. Up until this time, Eldorado had only been offered as a convertible. One can distinguish the 1956 Eldorado from the 1955 by the presence of the twin vertical wing hood ornaments used in the 1956 model year.

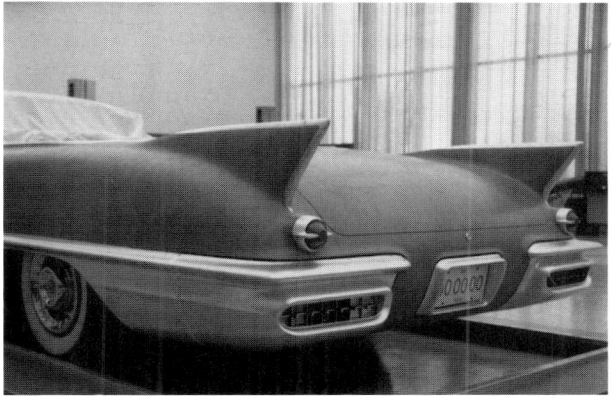

1958 Eldorado Biarritz clay model. Quickly one sees, even with this minimal facelift, that some of the drama of the 1957 has been lost in the new treatment of the rear bumper. With minor changes, this will be the form of the 1958 production model.

1958 Eldorado Biarritz fiberglass model. Luckily, by production time, the side scoops just aft and below the headlights were diminished. Notice that the vertical bars just before the rear wheels were enlarged, but reduced in number, for production. Also, the script was removed from the front fender just below the horizontal chrome strip. Though the texture of the grille changed, the amount of rubber on the Dagmars remained the same. Incidentally, the first man in the group behind the car is Bill Mitchell.

gold set on brushed chrome) were all that set the Eldorado off from the regular line. This was quite a change from the relatively austere 1953 concept, but one must remember that it was less expensive, and it was on this all-new body with those sexy hooded headlights. The interior leather was available in eight color combinations.

A More Beautiful Year

For 1955, the buyer got a completely new and distinctive Eldorado Special convertible. The new tailfin was inspired by Dave Holls' El Camino and was even placed above a similar accent spear on the rear fender that ended in the taillights. His patented chrome line running from the front bumper to the vertical upsweep just past the door lengthened the car visually. Even the engine had a full 20hp beyond the previous year and now included dual four-barrel carburetors. The sabre-spoke wheels, exclusive interiors, and vertical chrome bar lines over the rear bumper emphasize the car's individuality and sportiness. *Motor Trend* magazine tested the beast and deemed it a hot-performing car, maybe the hottest of the year. The 1955 Eldorado was indeed a special car.

Diversity in Luxury

For 1956, Cadillac decided to produce both a convertible and a hardtop, to be called the Biarritz and the Seville, respectively. This was in spite of the fact that DeSoto was also

1957 Eldorado Biarritz bubbletop built by Cadillac for a visit by Queen Elizabeth II. Constructed for Her Highness by General Motors of Canada, this was one of three specially prepared automobiles made available to the Queen when she arrived in Canada for her 1957 Royal Tour. The flag on the right front fender is the royal standard required by protocol.

using the Seville name that year. The Vicodec fabric top covering set the Seville apart from the rest of the Cadillac hardtop line-up. A dual-fin hood ornament replaced the old 1955 Cadillac goddess. From the rear, one can tell instantly the 1956 Eldorado from the 1955 because the 1956 has hooded chrome exhaust outlets with five vertical ridges and no vertical chrome bars over the top of the bumper. The finer front grille is also distinct from the 1955 and, since special plating was an option in 1956, some grilles are gold. Both Eldorados had 305hp engines and dual four-barrel carburetors, and they were the only Cadillacs that year not to have fender skirts.

For 1957, Cadillac expanded its number of Eldorado models. Just the year before, it had introduced the hardtop Seville, and now it introduced the very expensive Eldorado Brougham. Suddenly the Eldorado Biarritz and Eldorado Seville were regular or standard Eldorados! The development of the Eldorado line had moved a long way from Harley Earl's original idea to just build something different.

For the 1956 Motorama, Earl had two Eldorado Brougham show cars built. One was the formal Town Car with an open roof over the chauffeur's seat, and the other was the four-door pillarless hardtop that would actually be placed in production. The latter car was actually designed and built within a span of ten months, first being exhibited at the Waldorf Astoria Hotel in January 1955. Earl could hardly control his enthusiasm for the car and even hinted on television, in a rare appearance on Arthur Godfrey's show, that the Brougham might be put in limited production. Perhaps he felt he had to respond to Ford's growing string of hype over the Continental Mark II, but he made it more and more obvious over the ensuing months to whoever would listen that this was more than a "dream" car.

Enter the Brougham

Many designers consider 1957 to be one of Cadillac's high points. In 1957 Cadillac was presenting yet another all-new Eldorado design to the world. And, to a limited few who had the money, they were making available a totally new level of automotive excellence—the Brougham. The new Continental Division of the Ford Motor Company had introduced the Continental Mark II the previous year for the astonishingly high price (for the time) of $10,000. General Motors had to answer that challenge. They did it with the $13,074 Brougham. Both car companies lost money on every one of these cars they sold, but company pride was at stake. The Eldorado Brougham stretched the technological envelope as far as it would go for that time. The Mark II was stately, but the Brougham was state-of-the-art.

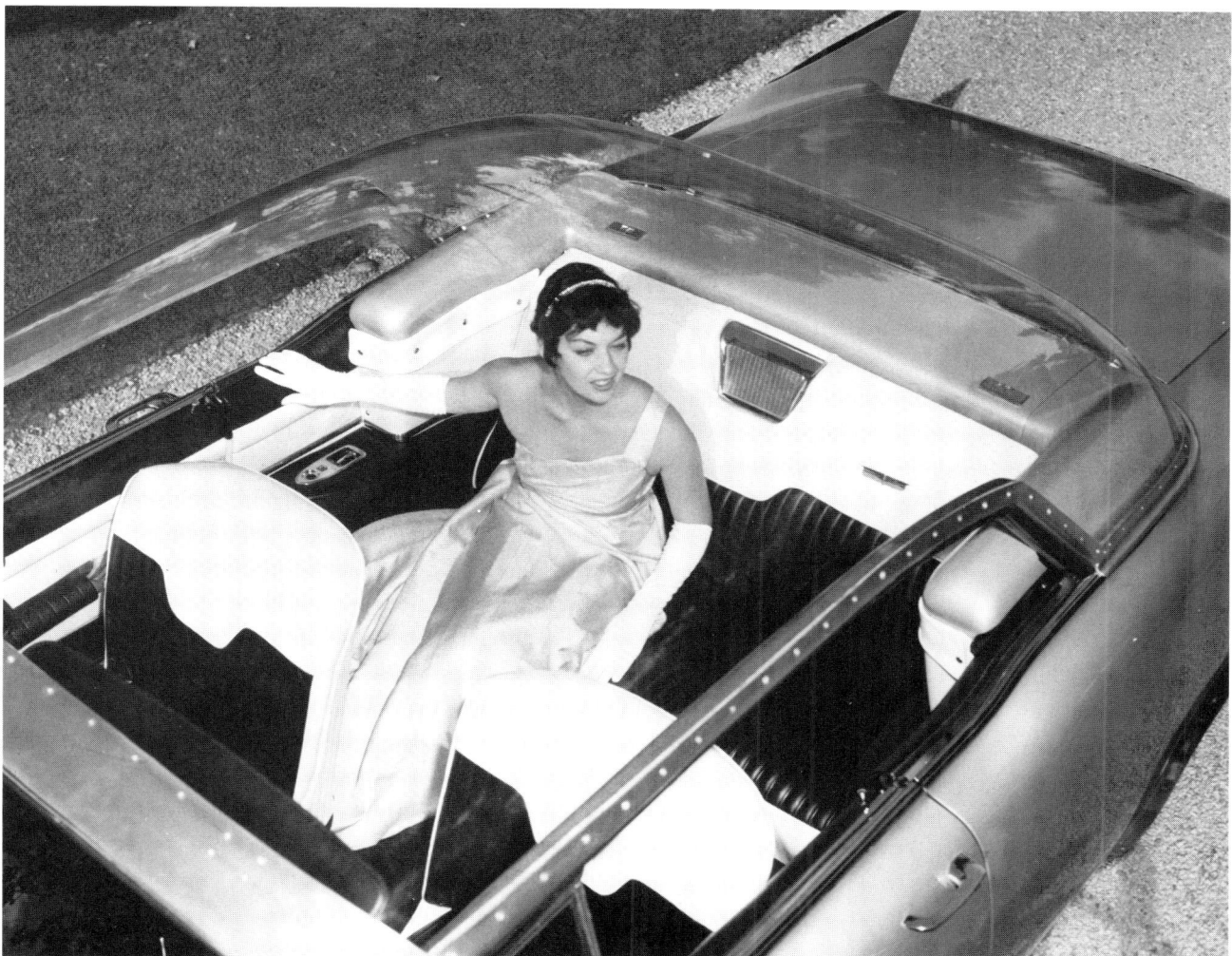
Even if it rained, Queen Elizabeth II's subjects could clearly see her through this transparent top. If the weather was clear, and the Queen approved, the top could quickly be removed for an unobstructed view.

Comparison pictures of the two cars, the Mark II and the Brougham, were the first photographs made on the turntable on the patio of GM's new Tech Center in Warren, Michigan.

By any standard, the "regular" Eldorados of 1957 were imposing automobiles. Rear ends are cheaper to redesign than front ends. If you start monkeying around with anything touching the cowl, things are likely to get expensive fast. Thus, the changes—and they were big ones—made on the Eldorado for 1957 were mostly done at the rear end.

Ron Hill is credited with designing the new rear end with the inset stabilizer tailfins. Hill was a fan of European sports cars and an advocate of smoothness and simplicity of line. His rear-end design is certainly less busy than that of the previous year. The cars are lower by three inches, and this low silhouette is enhanced by the sweep of the heavy chrome line beginning at the front of the skirtless rear wheelwell and terminating as the split bumper line dives under the car just before the license plate frame. All the rear lights are in the bumper halves and accentuate the side lines of the car. A 300hp engine was standard, with a 325hp dual-carb version optional. A total of 3,900 standard Eldorados were built, 1,800 convertibles (the Biarritz) and 2,100 hardtops (the Seville), at a list price of $6,648 for either model. The Series 62 convertible sold for $5,229 that year.

Only 400 of the virtually hand-built Broughams were offered during this its first year. These pillarless, four-door, stainless steel hardtopped cars marked Cadillac's reentry into the ultra-luxury automobile market for the first time since the production of their last V-16 seventeen years earlier. If stodgy stateliness was the problem of the Mark II, the complexity of the beautifully crafted Broughams was both its sparkle and its dark side. Yet, these cars were so far ahead of anything else on the road that one could learn to live with a little difficulty or

1957 Eldorado Biarritz white convertible. An impressive sight in your rearview mirror, you can almost feel the 300 horses under that massive hood. An optional 325hp engine was available that year for the Eldorado Biarritz and Seville. Note that the double wing ornament on the two fenders is similar to the single double-wing ornament seen on the 1956. Right at home in the hills of California, this beautiful example of the marque is driven often and well cared for by owner Joni Masket.

1957 Eldorado Biarritz convertible interior. This luxurious leather interior is complemented by crests and distinctive styling. That grille in the center below the dash is the speaker compartment. Owner: Joni Masket.

1957 Eldorado Biarritz convertible rear interior. Chromed Cadillac medallions are in the center of the two sides of the rear seat back, complemented by the speaker grille in the center. The solid tonneau cover provides an unobstructed view out the back. Owner: Joni Masket.

1958 Eldorado Biarritz convertible. For 1958, the Biarritz was still beautiful, but it had lost a little of its design clarity. Vertical accent bars appeared to the sides of the rear license plate frame and just before the rear wheels. Additionally, that clean sweep of mass behind the rear wheels to the base of the bumper was now changed and did not look as natural and sharp as in 1957. Note the "V" on the driver's side of the rear deck. Not seen here are the new quad headlights. This was still an exceptional car and it had to be. Cadillac was almost competing with itself because of the Brougham and the fact that the Series 62 now had so many of the design themes of the earlier 1956 Eldorado.

1958 Eldorado Biarritz fiberglass model with 1957 Eldorado Biarritz in the background. This is a comparison photograph made on the patio at the GM Tech Center when the 1958 Biarritz facelift was being prepared in October 1956. The vertical bars just before the rear wheel are much lower than on the final production model.

arrange to get some worrisome component modified.

Consider the complexity: there was no room left under the hood for the battery, so it had to be mounted in the trunk. The Brougham incorporated both the first use of GM's new air suspension, and Dan Adams' his-and-hers-memory-seat position mechanism. It was the industry's first dual-headlight model (though Nash and others produced a car with dual headlights that year, the Brougham's had evolved from GM's Motorama cars). The air suspension system proved to be the Achilles heel of these fine cars, even though GM had had extensive experience with such systems as used in its bus division since 1952. The ride was fantastic, despite *Motor Trend*'s lack of enthusiasm for it in their review. Passengers felt no impact from bumps encountered on the road.

The suspension's problem became apparent to me the first time I photographed one. The collectors whose

1958 Eldorado Seville. From the front, the studded front grille and quad headlights distinguishes this car from the 1957. Totally restored by Dwayne Medley of Texas, the automobile has won several national auto meet awards. Medley is a master craftsman. This car was not simply given a polish and trailered to a meet. He totally rebuilt the drivetrain himself and had to replace the top because the original was so badly rusted. Medley also had to rebuild the wheels of this car, learning as much as he could about alloys and the proper temperatures at which to work them. Today the wheels and center medallions look brand new.

Brougham I was to photograph had several cars and an elaborate complex of garages. One collector told me he would go get the car while I chose a place for the pictures. After a few minutes I heard the sound of compressed air. I thought someone was painting a car in another garage. It turned out that the man who had gone to get the car was putting air in the collapsed suspension. Air chronically leaked from the valves and connectors in the system, he said, and the car would not hold air for more than a couple of hours. He and his partner planned to replace it with a conventional suspension. Perhaps ironically, Cadillac had had the foresight to put a suspension system low-air-pressure warning light in front of the driver.

The car was a banquet of interesting features, including the incredible standard mouton or karakul carpeting, adjustable front seats (the seat traveled down and back automatically when the driver's door was open and returned to the preset position when the door was shut), and a trunk lid that could not only be opened, but closed from within the car. All doors locked automatically when the car started and if any door was open, the transmission selector could not be placed in any drive gear. The car was equipped with a completely stocked vanity case, which included an ounce of expensive perfume, a lady's compact, and a set of magnetized silver drinking cups. The radio's antenna automatically rose to "city height" when the radio was on and descended when it was turned off. Essentially, every option available on any car in the division—plus several that were not—were standard on this top-of-the-line Cadillac.

The X-frame developed for the car's special suspension was quickly adapted for use with the rest of the Cadillac line because it allowed for a combination of rigid support and low profile. Though the actual design of the 1957 Eldorado Brougham came primarily from Bob Scheelk, Harley Earl himself had developed and pushed through the ideas behind the car since the early days of Motorama. Even though it is said the Brougham was GM's answer to the Mark II, it was not introduced until about the time the Mark was taken out of production.

There were many differences between the Ford and GM offerings. The Mark had much hoopla preceding its introduction. The Brougham arrived on the scene with almost no fanfare. The Mark was a two-door with pretty mundane technological components. The Brougham was a pillarless

1955 Cadillac Eldorado Brougham (Seville) Show Car and Prototype. Appearing at Motorama in 1955, this four-door hardtop with quad headlights tested the water for the production Brougham program already underway. Note that there were no rubber tips on the Dagmar bumper ends. The tailfin of the car was substantially different—being more squared off and incorporating a taillight—from what went into production.

1956 Cadillac Eldorado Brougham Town Car. This show car was fiberglass and huge—219.9in long on a 129.5in wheelbase. The design sought to have things both ways: a space age design with an old-fashioned separate town car-style chauffeur's seat. Luxury appointments included a bar (with thermos bottle and glasses), radiotelephone, air conditioning, electric door locks, cigar humidor, and a lady's vanity unit. The brightwork in the passenger compartment was done in satin gold finish.

four-door technological marvel on wheels. Neither sold well, but the Eldorado Brougham, it could be argued, had more of a lasting effect on what its manufacturer, namely Cadillac, could and would do in the future.

The facelift for the 1958 Brougham only involved a change in engine carburation from two four-barrel carbs to three two-barrel carbs.

Nineteen fifty-eight marked a change for the car's primary designer, too, as he became head of the Cadillac interior studio after completing the 1957 Brougham.

Less than a Brougham, More than a Cadillac

If you think about it, Cadillac really didn't need the Eldorado Brougham to match the Continental Mark II. They already had the Eldorado Biarritz and the Eldorado Seville. Both of these cars were special-bodied Cadillacs with advanced styling and engineering. True, their front end was pretty much stock, but the profile and rear end set them apart from the rest of the Cadillac line. For 1957 and 1958, 3,904 copies of the standard Eldorados were sold, compared with just 3,000 Marks for its total production.

The Italian Connection

The 1959 and 1960 Eldorado Brougham program was one of the most unusual in GM history and a complete departure from what had been tried before. General Motors reached an agreement with the Italian coachbuilding company, Pininfarina, to build the specially designed Eldorado Broughams for these two years. Battista Pininfarina had founded this renowned carrozzeria in Torino in 1930 and had built bodies for an incredible array of automobile manufacturers, including Ferrari. The founder's son-in-law, Renzo Carli, ran the company from 1958 when Battista retired until Sergio Pininfarina took the helm in 1966. The idea was that the Broughams would be designed by GM and their chassis and engines manufactured and assembled in the United States. These would then be shipped to Italy for the construction of the bodies and interiors by Pininfarina. (Chrysler had had a similar arrangement with Ghia for the construction of their Imperial limousines.)

Sometimes these Broughams are mistaken for the 1961 and 1962 Cadillac sedans. They were certainly not the flashy splashy Broughams of 1957 and 1958! Dave Holls designed the crisp formal greenhouse on these cars, and the narrower sail panel of the Brougham is the first hint that you're not looking at a 1961. As you look at the car awhile longer, you appreciate the understated elegance of this hand-built body. These are magnificent automobiles and, partially because of their superficial similarity to later standard Cadillacs and their low production numbers, they have been largely overlooked by collectors. The surviving numbers are lower still because, being handmade, a lot of lead, red oxide, and other fillers were used in the bodies, making them difficult to maintain.

Compared to the other cars in the Cadillac 1959 and 1960 line, everything about this model was understated. They shared the 130 inch wheelbase with the rest of the division, but the bullet taillights of 1959 were missing, the fin size was reduced, and the windshield was not the severe wraparound type used in the rest of the Cadillacs.

Similarities with the 1957 and 1958 Broughams are few. They did have a redesigned air suspension system, but

1956 Cadillac Eldorado Brougham Town Car (show car). In this GM publicity photo, the elegant machine is shown in service at the front door of a stylish (for the fifties) estate.

little niceties such as the silver cup set were no more.

The Pininfarina badge appeared on the 1959 model, but not the 1960. This was understandable, since the cars were designed in the United States. As a rule, Pininfarina applies its badge only to those cars it designs, such as the more recent Allanté, also by Cadillac. The 1960 Brougham had many styling similarities in common with the 1961 and 1962 Cadillac line.

Variation on a Theme

The Eldorado Biarritz and Eldorado Seville for 1959 and 1960 were harbingers of what Eldorado was going to be until 1967—a gussied-up standard bodystyle Cadillac noted for its differences in exterior chrome, standard equipment, trim, and its fancier interior. This in no way belittles the 1959 and 1960 models—these cars will never be forgotten. Dave Holls was the primary designer on the 1959 exterior and Bob Scheelk the primary designer on the interior. The totality of their design effort has so much presence that the 1959 and 1960 Cadillacs have almost become American cultural icons. (Originally, the 1959 Cadillac had been planned as a more subdued car than it turned out to be. However, during the design process, some of the Cadillac designers saw Chrysler's big-finned new cars in a nearby factory parking lot. Suddenly, it was as if a challenge had been made, and Cadillac jumped into a design program featuring big fins.) But for a time, Cadillac Eldorado was to move away from the striking individualistic design themes of the fifties to the more elegant and subdued appearance of the early 1960s.

This was, in part, due to personnel changes. The dynamic and authoritarian Harley Earl, who had created GM design out of nothing, retired in 1958. He had been the king of flash and had had a good eye for what America wanted through most of his career. In fact, there would never have been a single Eldorado had it not been for his genius. But the overchromed 1958 GM model line-up was his swan song, and many designers, even though they loved the man and what he stood for, agree that he retired just in time.

Bill Mitchell replaced Harley Earl as Vice President of Design. Ed Glowacke continued as Chief Designer at Cadillac Studio until the time that the 1960 facelift began. At about that time, Chuck Jordan took Glowacke's position and Glowacke became Bill Mitchell's assistant.

Chuck Jordan oversaw the final details on the facelift for the 1960 Cadillac line before beginning his program of trying to restore the elegance and prestige he thought the marque had lost with the excesses of the late fifties.

Bob Scheelk Interview
Eldorados of the 1950s and 1960s

Bob Scheelk studied design through a correspondence course before joining General Motors in 1947. He is acknowledged as the primary designer of the 1957 and 1958 Cadillac Eldorado Brougham. After that time, Scheelk was responsible for Cadillac interior design until 1965.

Author: You joined Cadillac Studio in '53. Were you involved in the '54 facelift at all?
BS: The '54 was a whole new car from the ground up. I went from Oldsmobile into Cadillac in February of 1953. But the '54s were all pretty much wrapped up. I mean, they had a complete interior/exterior model. For all practical purposes, it was done.

Author: Well, what about the '55 then?
BS: That was a facelift of the '54. It entailed a new bumper and grille and a new side treatment. The side treatment for the car was done by Dave Holls.

Author: Yes, he said he had a patent on that.
BS: Yes. It was the chrome molding starting from the front that ran to the rear quarter, took a ninety-degree turn, and went up to the top of the fender. The Eldorado received a new rear-end treatment, including fenders, taillights, and bumper ends. Also, it was equipped with aluminum wheels.

Author: What about the next year, the '56?
BS: The '56? Another facelift program. I did the instrument panel for the '56 which was new and I was pretty much kept busy with that. I remember the front end seemed to have evolved more or less during a late night session in the auditorium under the direct supervision of Mr. Earl. I wasn't in on it. They widened the hood and revamped the grille and front bumpers. The rear sheet metal of the Eldorado was not changed.

Author: And, Ron Hill did the rear end on the '57?
BS: Yes. That was a whole new program.

Author: Were your—and I mean *all* of them—designers working on that project at the time, getting ideas for the rear end taken from the European cars?
BS: No, we were just playing around with a different approach to the fins and to the rear, in general. That's all. We were just searching. We weren't influenced by anything going on in Europe that I can remember. It was just an attempt to get a different looking rear end.

The Cadillac that I really was more involved with was the '57 and '58 Eldorado Brougham. That car was responsible for a number of patents for me. That was a nice courtesy thing—a form of recognition.

I think it would have been better if we had been able to do the basic series of Cadillacs first and then the Eldorado Brougham, but it didn't work out that way. We had to design a basic Cadillac that was still a Cadillac, yet didn't compete with the Brougham, designwise, as well as costwise.

Author: Do you think that's why the later '58 regular Cadillac ended up looking like it had so much chrome in it?
BS: No, I think it was just a trend of the times and, as far as that goes, I think the '58 Cadillac was relatively clean compared to a lot of other cars. It isn't necessarily how much chrome, but how well it is distributed. But understand, I am not an advocate of a lot of chrome.

Bob Scheelk.

Author: Do you have any drawings from that sequence?
BS: As far as sketches, I had a few that dealt with the front, rear, and side treatments, but they were donated to the Edison Institute in Greenfield Village sometime during the 1980s.

Author: What I don't understand is, what was sent to Italy when Pininfarina did the body for the '59 and '60?
BS: For Italy? Yes, that was the '59 and '60 Eldorado Brougham. That was the one that was sent to Pininfarina to be built. By that time I had the Cadillac Interior Studio so we designed the interior for those cars, but they were trimmed in Italy.

Author: The '57 and '58 Broughams were done in the United States, correct?
BS: Yes. And the '59 and '60 were designed in the United States and fabricated in Italy. In fact, I think it was primarily the body. I think there were a lot of components for the car that were shipped to Italy from the United States and assembled in Italy by Pininfarina. I'm not in a position to say what was, or was not, shipped to Italy.

Author: The '57 Eldorado Brougham that you designed had four doors, and the rear doors open backward.
BS: Yes. It had a short stub pillar or lock pillar at the center and, by hinging the rear doors at the rear, it facilitated entrance room into what was a relatively close-coupled car.

Author: So who headed up the team as far as the '57 was concerned?
BS: Well, Glowacke was the studio chief designer, I was acting assistant, and Frank Biondo was the studio engineer. At the time, Hill was still in the studio and a fellow by the name of Ned Walters. Herbie Kadeau was in there when I came in, but he got reassigned to work for Mitchell in some capacity. Jens Moltrud was the chief modeler, and he was a good one. That was the setup.

Author: Was Glowacke actively involved in the design or just giving advice?
BS: He was approving things or disapproving things and guiding our efforts. Of course, we were seldom without the attention of Mr. Earl. I can't remember the exchanges that took place or what modifications may have been made as a result of Ed Glowacke or Harley Earl, but I know that I was quite active on that car and I contributed a lot to it. So did Henry Lauve, Marcel DeVoss, and Fidele Bianco, to the interior.

Author: The '57 Eldorado, the one that's not a Brougham, how much work did you do on that car? Was that another one that was cut down from the Brougham?
BS: You're talking about the '57 Eldorado convertible? Well, that was a brand-new body. It wasn't carried over from '56 at all, and it spun off of the regular Coupe DeVille and convertible, except that it had entirely different rear quarters and interior trim. The front end was the same as the regular production car except for a couple of ornaments on the fenders, etc. I was more involved with the front end.

Author: Who did the rear on that one?
BS: That's the one that Mr. Hill did, for '57 and '58.

Author: Now, you guys did not call those front bumpers "Dagmars," you called them "bombs," correct?
BS: I'm trying to think how we did refer to them. I can't remember any particular name.

Author: Mitchell told me a long time ago that Glowacke had done those the first time.
BS: He should know. I think they first showed up in production about 1950 or 1951. I think he was the studio head at the time.

Author: You changed from exterior to interior during this time?
BS: Yes. I wasn't in on the exterior starting with the '59 models. By that time I had the Cadillac Interior Studio.

1957 Eldorado Brougham fiberglass model. Using sabre-spoke wheels like those of the production 1955 Eldorado, this August 1955 model was close to where the designers wanted the lines to be for production. Note that the model does not have quad headlights, and there are no rubber tips on those massive Dagmars. This photograph—with the car on a turntable built into a linoleum tile floor—was taken at the time the design studios were moved from downtown Detroit to the Tech Center in Warren, Michigan. This photo was most likely done in the old facility.

Our studio had the '58 interiors to do, and then brand-new interiors for '59 and '60. I wasn't involved with the exterior at all after the 1958 models.

Author: When you were doing the '59 and '60, is there a special character you tried to give to Eldorado that wasn't in the regular line of Cadillacs?

BS: Are you speaking of the Brougham or the regular Eldorado convertible and coupe interiors?

Author: Well, both of them with Eldorado names on them.

BS: Oh, yes. We did quite a lot of work on the Eldorado, both the convertible [the Biarritz], the coupe [the Seville], and the Brougham itself. We went to special lengths on the interior. I think the interior came off very nicely. Not many of them are still around. We did an extra-special effort for the Eldorado Brougham that was produced in Italy, and also for the other Eldorados in the American car line.

Author: I asked another designer if he had an Eldorado during that time and he just laughed. He said he couldn't have afforded one. He said that when Earl found out how much the '53 was costing he went through the roof. What kind of cars did you drive back then? Holls had a Corvette and Hill had a Porsche.

BS: I came into Cadillac with a '53 Oldsmobile and then I got my first Cadillac. It was a '54 Coupe DeVille. I had a slew of them over a slew of years. I had a 1968 Eldorado which I think was a very handsome car.

Author: What was the last Eldorado on which you worked?

BS: The '65 program was the last one that I did as head of the Cadillac Interior Studio. Then I was made an assistant to Mr. McDaniel, and George Moon took over the Cadillac Interior Studio. He got in on the '66 program.

1957 Cadillac Eldorado Brougham, interior design buck. This mockup shows a near final-stage version of the design for the lush Brougham interior. Note the sea of luxurious mouton carpeting. It makes your feet feel good just to look at it.

Ron Hill Interview
Eldorado in the 1950s

Ron Hill worked in Cadillac Studio as a designer in the 1950s. after decades at GM, he returned to where he had received his training, the Art Center of Pasadena in California, as a design professor.

Author: You are well known for your contribution to the 1957 Eldorado. Could you say something about your experiences with that project?

RH: We started working on a variety of projects, that being just one of them, and ended up with the 1957 Cadillac program which started later in that year. And that was a very interesting program; the first full car line that I ever worked on because prior to that we just did facelift work on '56 Cadillacs and what have you. But the '57 was an all-new car, as you know. And it just so happened that we got involved to the point where we were doing various things and the one thing that we were doing was looking for a distinctive appearance for the Eldorado, specifically the rear quarters—the rear end of the vehicle. And there was money to spend on that, so we just submitted a variety of different ideas and somehow mine prevailed.

Author: Speaking of money, though, were there any real constraints?

RH: Not many. It had to use the body forward of the doors and the front end of the standard '57 model which, of course, we all worked on also.

Author: So you could use the front doors? But, as I understand it, the Eldorado Brougham had doors that opened from the back. What the collectors call the "suicide door."

RH: That was an entirely different program. That was one that was based on a show car. That kind of was a separate program—a separate stream if you will. Separate, but parallel.

Author: I know that, but the reason I brought it up was this: How much of the flavor did the designer of the Brougham take from the rear-end treatment that you developed?

RH: Actually the Brougham was developed substantially before. That had been in process for some time. And that heavily influenced the '57 line. But stylistically it did not influence the Biarritz Eldorado at all because, of course, we were looking for something entirely different.

Ron Hill.

Author: Who were the other guys working with you?

RH: Bob Scheelk, I think, at the time. Bob, I believe, was the assistant designer [assistant to Glowacke] at the time. Oh golly, who else was in there?

Author: Did Scheelk work mostly on the interior, or did he also work on the exterior?

RH: He was working on the exterior at that time. He was the assistant chief designer. Then after a while he moved somewhere else and I became the assistant chief designer, but I believe that was after the '57 program.

Author: What was Bill Mitchell doing?

RH: Mitchell was, of course, director of styling at the time under Harley Earl. Harley Earl had something to do with the program, but not an awful lot. It was mainly Bill and Jules Andrade and Ed Glowacke, who was the studio chief at that time.

Author: Did Mitchell and Andrade actually have any input into the design, or did they just make suggestions?

RH: Just made suggestions and accepted things. That kind of thing. It was primarily that the rest of the people in the studio—and I'm trying to remember who the devil they were—but, at any rate, it was a small group; there were only three or four of us as designers, and then Bob Scheelk was the assistant and Ed Glowacke was the studio chief, but Jules and, especially, Mitchell were management contacts. Harley Earl rarely . . . I mean, he very much exerted an influence, but he wasn't that much of a design force on the '57; he got more involved in the '58 and then, as I understand it, because I went into the Army, he dropped out entirely after the beginning of the '59 program and from there he retired at the beginning of 1958, I think it was.

Author: Mitchell told me that Glowacke was the one responsible for the Dagmar bumpers. Is that your understanding?

RH: Oh, that's quite possibly true. They existed before I got into the studio. I mean, we had "the bombs" as we called them. We rarely called them Dagmars—"the bombs." I don't know why. Sort of when I got there in January '55 they were de facto. And what we did in '57, of

1957 Eldorado Brougham prototype with Lincoln Continental Mark II, a comparison. The big competitor for the Brougham was supposed to have been the Mark II, but it went out of production by the time the Brougham went on sale. No one knew that would happen in November 1955 when this photograph was taken by the turntable on the patio of the GM Tech Center. This picture clearly shows what Dave Holls meant when he said, "One car on that patio looked like it was from Mars and one looked like the Duncan Phyfe factory made it." Holls does say that the Mark II design has worn well over time, but the advanced design of the Brougham is readily apparent in this comparison. The large structure in the back is the styling dome. Dave Holls believes these to be the first pictures ever taken on the new GM Tech Center patio, a space specifically designed by noted architect Eero Saarinen, to show and photograph automobile designs. Saarinen's 900-acre design and technical complex included a twenty-two-acre lake and other sweeping vistas suitable for automotive photography. (Retired GM designer George Moon is writing a history of the Tech Center.)

course, was put the rubber nipples on them so that they wouldn't be perceived as being quite as aggressive. But they were very much a stylistic force from probably about . . . they must have been designed sometime around '53, I would guess.

Author: Do you remember the process that you went through in developing that rear end?
RH: Yeah. At that time the idea was that you had young talent to give you ideas and show you a new, shining path to absolute competitiveness and domination of the marketplace. And we would try just about anything. You know, that's what we were asked to do, so we came up with a variety of ideas and I did the somewhat radical thing of trying to do a complete smooth form, then superimposing the fins on them—rather than having them an extension or a growth out of the side form where it meets the rear-end form or the upper form. And, to do that, that was considered fairly radical at the time. I used to get razzed a lot, particularly by the people who really knew what they were doing—which was the modeling crew. In fact, I had a bet with the assistant chief modeler that the design would never fly. I said, "OK, I'll take you on." I won what was the ultimate prize in those days—a bottle of whiskey! That was the year of "drink a lot and do bad cars." Quite literally. And I was just this dumb young kid out of California who didn't know any better.

Author: How much interaction did you have with the other guys who were working on the front, the sides, and the rest of the exterior of the '57?
RH: Those were pretty well set from the standard car. What you did was establish a theme for the vehicle, which was really based on the show car which later became the Brougham—the vehicle with the suicide doors that you remarked on. That sort of established the theme of the vehicle but, of course, the fins were canted forward instead of back and there were a lot of detail differences. That was a very interesting program because it went from basically a forties and thirties style influence to something that was radically different.

Author: What about Dave Holls?
RH: Dave came into that program quite a bit later. I'm not sure he was in the studio at that time because he had come back from the Army, and then came back in the

The extreme difference in overall height is readily apparent in this shot of the rear of the two cars. The Brougham was 54in tall, and the Mark II, 56in—quite enough to give it an old-style look. Certainly, the hump for the rear tire in the rear of the Mark II does not yield a youthful appearance in comparison with the Brougham.

studio. I'm not sure about the timing. That is, don't forget, forty years ago.

Author: I know he worked on the '59.
RH: Yes, he did. I did not, but he did. And I was in the Army at that time. Actually, we had just started the '59 program when I was drafted and I came back after. I left in November of '56 to be drafted and came back in two years, January '59. That's when I returned and spent some time in another studio and eventually ended up back in Cadillac after a year or two. About '60 or so. I know it was '60 because we were just finishing the '61 program. I worked on the '62 when I left Cadillac, and I never worked in Cadillac again.

Author: Oh, so you worked on the '62? Were you working on the rear end, too? That's another very distinctive rear end.
RH: On the boot. That was one of my designs, yes. That boot or the exhaust combination boot and backup lamp. That sort of thing. That was my last encounter with Cadillac. I never worked with them after that. I spent many years with Pontiac and Chevrolet and Buick and overseas, but never worked in Cadillac again, and I never worked for Oldsmobile, which is very curious. Oldsmobile or Holden. But I worked for GM Brazil and Opel and Vauxhall. And of course, the legendary Harley Earl. Particularly, Harley Earl because he was such a charismatic figure and you never knew what he was going to do or say and it was just ultimate strangeness to us young designers.

Author: What about Ed Glowacke?
RH: Ed was actually a very good boss. I enjoyed Ed a lot. He did a lot for me. He was very supportive. He liked me and I liked him and we had a very good relationship. He promoted me to assistant in there, which was very nice for a twenty-two- or twenty-three-year-old kid.

Author: I know he died fairly young.
RH: He died in 1962 and I remember that because I had taken a late honeymoon, got married in October '61, and then in June '62 my wife and I went to Europe. Something I wanted to do after having been there in the Army, and that was the point where he died. He would have been, no question about it, he would have been the next head of GM Design after Mitchell retired.

Author: I know Mitchell always spoke very highly of him.
RH: Ed was on a very fast rising course in his career.

Author: Well he was an unusual person for the time. He did sky diving, flying, and racing.
RH: Yeah, he had a pilot's license. He used to go flying with us and he raced cars and everything. He was an interesting man. Apparently he had a problem with his health and realized that it was a chancy thing, and I think

Above and below, 1957 Eldorado Brougham fiberglass model. These two photos, taken in June 1955 compare the front end design of the Brougham with the quad versus two headlight configurations. Clearly, the massive crest of the hooded fender was just too much to use the two-headlight configuration and quads were used in the production model.

that sort of put life on edge for him. He used to talk to me about that occasionally—and I didn't fully understand what he was getting at the time. Of course, I do now.

Author: What do you think that General Motors wanted the Eldorado to be?

RH: To be distinctive, above all. They wanted a person that was going to spend the extra money to have this rather flamboyant vehicle—have something that was quite unique and distinctive. That was our understanding, because we tried to do that, of course, with everything we did. But here was a chance when you took basically a production car and had some opportunity and the luxury of doing some exploration and moneys to spend on primarily the rear quarters to give the vehicle some distinction. We tried all kinds of different things and I just happened to come up with a combination that struck a cord at that time and, you know, a lot of it's luck and how well you sell it and a good sketch and a boss who respects you. Those things can happen. Although, there's been probably many, many more good designs that have been passed over than have been used, but you just never know. I think about it today and I think it was a fantastic piece of good luck on my part. I have a model of it right in front of me now, a little scale model. In fact, my son got it for me. It's a French Solido. It's the first one I've seen of that particular car.

Author: Did you actually have an Eldorado later?
RH: No, I never had a Cadillac. I was never interested. I've never owned a Cadillac. I've owned many of the cars GM made, but for some reason I just never had a Cadillac.

Author: Did you meet any enthusiast Eldorado owners later?
RH: No, there was never any occasion to do so because it was merely a footnote in history. It was one year that the Eldorado existed in that condition, it was modified for the '58 run, and then of course in '59 it took an entirely different flip-flop. I think the '57 was unique because it was the first time we were looking at a form like that, three-dimensionally rather than two-dimensionally. In other words, rather than a side and a front, it was a total form. That, to me, was the crowning achievement. Not what it meant stylistically, because it was a large car, it was very ponderous, it was heavy. Certainly the concept would have worked much better on a smaller, less massive kind of form.

Author: Were you influenced by any of the experimental cars?
RH: Oh yeah. But actually, we were influenced by what we saw happening in Europe. To us, European auto design was the most interesting thing. I'm talking about the young people, the young designers. We didn't believe in "the juke box" sort of thing that was going on. We sort of went along with it because that was what we were getting paid to do, but most of us were either getting or had sports cars.

Author: Which cars from Europe are you talking about?
RH: Oh, just about any. Particularly the Italian cars. But almost all of them. The English were fairly dominant at that time. In fact, when I got out of the Army and came back and worked for about a year, I had a Porsche and that was very unusual at that time.

Author: So, what made you decide to go back to Art Center?
RH: After thirty-one years and going through a variety of different areas and, you know, having a fair amount of interesting assignments and challenges and of course all the awkward things that you go through, I had an offer from the school. I had kept in touch with the school because I'm from this area and they made me an offer. Actually, it was hinted at in 1980 and then again in 1985. My predecessor's health deteriorated to the point where they had to do something, so they asked me. They made an offer, and I was able to negotiate through the good offices of Chuck Jordan and Herb Rybicki, who was VP of Design at GM at the time. They enabled me to make a fairly painless transition from the corporation to academia. It was time for a career change. I never thought I'd come back to California. It was a surprise to me in some ways. My parents are still living, which is, of course, an added benefit. But that's not the primary reason. I did it because it was time to make a change.

1957 Eldorado Brougham production model. The clean futuristic lines look just as good today as they did almost forty years ago. At first glance, this does not look like a four-door car, but it is. The car has the sporty appearance of a two-door hardtop and all the conveniences of a four-door. Only 400 were built for this model year at the considerable price (for the time) of $13,074. This was at a time when one could buy a new basic Cadillac convertible for a mere $5,229. Even with this price difference, GM lost money on every one of these practically hand-built cars, just as Ford did with the Mark II. Owners: Herb Rothman and Ted Davidson.

Author: Could you comment on the fact that many times, the divisions at GM have shared front doors—that because of this, the designer often has to start at the cowl or front door in developing a new design.

RH: That is very common because you could get a new look just by making a new set of tools for a rear quarter. It is much less expensive because don't forget, doors have things like glass that works; they have to have glass drop and all the mechanisms to support it, so that's a pretty expensive piece of real estate. To change the rear quarter—where you don't have the consideration of glass drop and always have things as a fixed piece, not hinged—it's a little less expensive piece of real estate.

Author: So that was a common order that went on for a few years? I say that because of the distinctive '57 Eldorado rear end you worked on.

RH: Yes, you have to understand that, at the time, Fisher Body did all of the body building in components. In other words, they would build the central part of the vehicle, that is, the passenger compartment, the roof, and the cowl. The cowl is the most expensive piece of real estate on a vehicle because everything comes together there. The heat vent, electrics, instrumentation, and the controls. It's that area the car is really built around—it determines the packaging and the placement of components.

Author: I guess the front doors are, in effect, mounted on the cowl.

RH: The front doors are mounted to the cowl and that's all part of it. Usually what the divisions would do was, of course, control the look of the instrument panel, but not the gubbins underneath. They would do all the front-end sheet metal—usually the suspension, running gear, and engine, that sort of thing. But that body in white was all Fisher Body at the time and we had what we called the body room. The people of the body room tried to rationalize because there were cost considerations even in those days, where you'd make a minor move on a bumper and it was going to cost the corporation a couple of million

1958 Eldorado Brougham. Here the incredibly well-engineered and strong abbreviated center pillar is visible beside the back of the front seat. The interior is luxurious and getting in and out is no problem. Including the roof, a great deal of bright stainless steel was used on the body of the car. Owner: Garth Higgins of California.

The price did not go up on these beauties for 1958, but the horsepower was increased to 335. Only 304 copies of this magnificent machine were made during this, the last year of production. Very few changes were made between the two model years. Owner: Alan Dowling, current President of the Brougham Owners Association, Inc.

dollars. Think how many times that would happen. But on the other hand, in order to remain competitive and have a fresh looking product, you had to make these kinds of moves. This was the juggling act. And, despite rumors to the contrary, it went on even then. There had to be some rationalization. One of the ways to rationalize was to hold the front doors standard for more than one division, and try to change everything either forward or aft of that.

Author: Who worked with you on the '62 Eldorado?
RH: Chuck Jordan was running the studio at that time and I was brought in as assistant. Ed Taylor, Jerry Brockstein, Don Roper, Mike Golden were some of the people who were also in the studio at that time. That's a little closer; a little easier to remember because I tend to put everything into pre-Army and post-Army—that was a definitive event in my life, you know, to be jerked out of a nice job with a bright career and to be thrown into field infantry.

Author: You traveled a bit in the Army as an illustrator. Did that affect your idea of design?

RH: Yes, indeed. It made a difference in my outlook and attitude when I came back. Everybody matures. Things do change. I think the Army and travel did influence me. And, of course, I think after that we saw more sophisticated designs. I think the '60s were a pretty good era for design and it got much better up to '68; then the realities of insurance, the realities of cost, and the realities of finite energy sources, plus pollution, came into being. The '70s were a very difficult era. It was an interesting challenge, but for people like Bill Mitchell it was very frustrating because he had finally achieved the plateau, the heights that he wanted. He finally had the power to do the kinds of cars he wanted to, but he was blocked from doing them because the marketplace and the public's perceptions just took a 180-degree turn or, at least, a 90-degree turn.

Author: You mean he was all dressed up and had nowhere to go.
RH: Exactly. It was a very frustrating era for the man and I can understand why. I didn't agree with it because I thought it was a great challenge. Something to be dealt with.

1957 Eldorado Brougham rear end. The curves and detailing in this highly sculptured design were exquisite. The forms designed into the bumpers required the use of aluminum instead of steel for their construction. Even the gas filler door was given careful attention to placement and craftsmanship. Owners: Herb Rothman and Ted Davidson.

Dan Adams Interview
Cadillac Engineering

Dan Adams joined GM in 1934 and stayed for forty-five years working with Cadillac. He solved many engineering problems for Cadillac and worked as a body engineer. Adams headed the section for future planning for Cadillac for two years before retiring in 1979.

Author: The big GM strike occurred between August 1945, and April 1946. For a time while that was going on, part of '48 Cadillac design was completed on Franklin Q. Hershey's farm outside Detroit.

DA: One thing we did was to form a company over in Chicago called the Catalina Motor Car Company. We hired draftsmen over there and used the Chicago branch as our working place. We designed and made the templates for the Keller models and so forth in Chicago. I had just gotten out of the service. Dreystadt [Cadillac General Manager at the time] was madder than hell at me. I had an airplane before the war and I kept hearing from my draft board, so I had enlisted in the Army Air Corp. I was in that for three years to September 1945.

When I returned to Cadillac, I was made assistant staff body engineer, sheet metal and body. We had to go to Chicago to do our work because of the picketing in the Detroit area. That's where we came up with the first fishtail mechanism that opened to fill the gas tank.

We were over there for several months. It seemed to me like a long time. I know because my son was born in December and I had to commute back and forth from Chicago.

I worked closely with styling and I had a lot in common with Bill Mitchell. He had been in the service, too. He was doing art work and so forth for them. He was making up posters for training and things like that.

Bill, of course, he'd always get a Cadillac and soup it up and I'd just get a regular Cadillac. Then we'd go off by the railroad tracks and see who could beat each other. He'd always spend a lot of money on it and I could still beat him with a regular Cadillac. Bill was a great guy.

You can tell his designs. He was very much a fastback guy. You know what I mean about the back end of the car. He'd do a fast sloping back on the car right down to the bumper. And there were several cars like that. In fact, Cadillac had a two-door that was made like that right after the war. And Buick had a very nice one.

Dan Adams.

Author: What was the first Eldorado you remember working on?

DA: The '54. We put extruded aluminum on the rear wheel opening cover and along the rocker panels. That was the first automotive body use of that material for this trim location. This was below the low belt.

I remember going down to Reynolds aluminum in Louisville to get them going on that project. At that time, I was body engineer for Cadillac.

I also was assigned to Cadillac's work with Pininfarina on two occasions.

Author: Did you go to Italy to work on the Eldorado Brougham project?

DA: Yes. I was very much involved in the Brougham project.

Author: What about the one that was built here—the '57?

DA: Well, I was involved.

Author: Can you tell me any special problems you had with the Brougham project?

DA: Well, getting the Brougham [1959 and 1960] built in Italy was a problem. The people in Italy really didn't have a timetable like we do, and *domani* was always good enough. The first car was supposed to be delivered in September. They said it would be ready in a week or so—and this went on for a couple of weeks—so I flew over there and they hadn't really gotten the thing sorted out. A fellow named John Doll was another engineer that we had. He went over there and helped them straighten out the parts. We shipped all the parts over there that would be used commonly. A simple thing like a nut would get screwed up because "nut" meant "tree" to them. But we finally got it straightened out and we built ninety-nine for '59 and 101 for '60.

We sold them [the 1957 and 1958 Eldorados] for around $13,000. The car had a lot of innovative things. It had no center pillar—a four-door sedan with no center pillar! It had a stump pillar that stuck up, that stood up to carry the locks. The front door hinged from the front, and the rear door hinged from the rear. When you pulled the lever up to open the rear door, the window automatically came down to make the clearance to open. There were a lot of completely new things on that car. I think the dealers got to know the owners of those cars personally. The stainless steel roof was an integral part of the structure on that car.

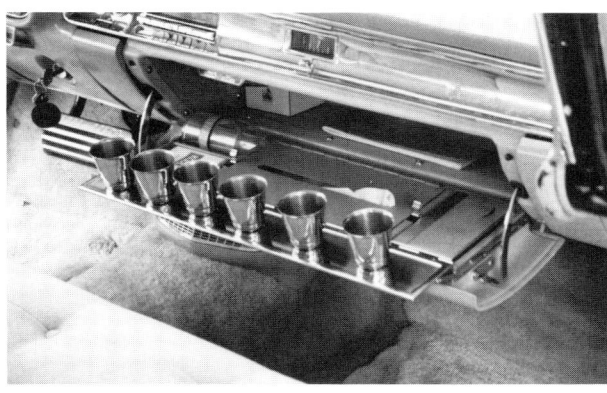

1958 Eldorado Brougham: the standard was luxury. The standard luxury components of the Brougham were legion. Here is the set of six magnetized, silver-finish drink cups which came in a clear, plastic container with the car. Also included were such things as a folding vanity mirror; a vanity compact containing a coin holder, powder and puff, lipstick, comb, mirror, loose cigarettes, custom-fitted cigarette case, and a built-in tissue dispenser. Owner: Alan Dowling.

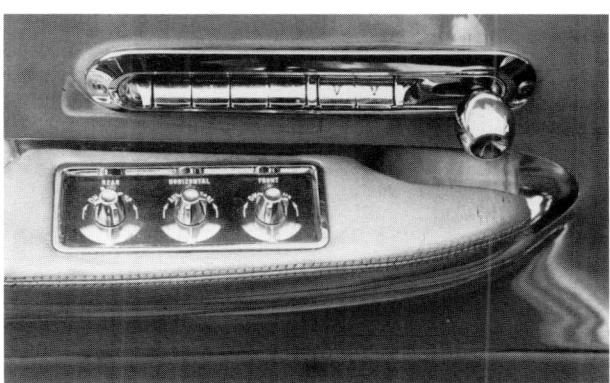

1957 Eldorado Brougham, "Favorite Position Seat." Designed and patented by GM engineer Dan Adams, with a colleague, this device worked well without benefit of such things as computer chips. The circuitry was based on training system design work that Adams had done in the Army Air Corps during World War II. Owners: Herb Rothman and Ted Davidson.

1957 Eldorado Brougham. Most luxury cars today have rear seat armrests, but this one opened to reveal a note pad with pencil, an atomizer containing *Arpege Extrait de Lanvin* perfume, and a portable, beveled glass vanity mirror. The note pad, mirror, and compact were bound in leather that matched the car's interior. At the base of the rear seats are two adjustable heater grilles and between the two rear seat positions is the radio speaker grille. Owners: Herb Rothman and Ted Davidson.

1957 Eldorado Brougham dashboard. The elegant instrument panel and dash provided easy access to accessories and controls for the driver. The plastic blade turn signal indicator lights just above each end of the horizontal speedometer are a dramatic and useful accent to the dash design. Owners: Herb Rothman and Ted Davidson.

Author: Dave Holls told me that there was a big problem painting those things in Pininfarina's booths because the Broughams were so long.
DA: Yes, but the Italians adapted pretty quickly and did a good job of it. The painting wasn't a real big problem.

Author: I heard that about this time Salvador Dali presented a design for Cadillac.

DA: Yes, at the auto show in Paris, they showed the Eldorado and the Brougham. It was Harlow Curtice's [GM President] wife, I believe, who had lunch with Salvador Dali. He convinced her that he would be a good designer for a Cadillac show car to be shown in the Paris Auto Show. Eventually, the powers-that-be got down to me to deal with this. I got a hold of this fellow from legal staff who I knew was of French descent and spoke some

1957 Eldorado Brougham. The rubber tips on the Dagmars soften the otherwise aggressive attitude of the front end that was seen in the all-metal Dagmars of the early prototype. The quad headlights are an integral part of the design, adding to the horizontal themes in the front of the car. The pancake-style hood was relatively expensive to produce and is front-hinged. The vents across the bottom of the windshield are functional air intakes for air conditioning, heating, and ventilation. The long louvered dual vents on the crowns of both front fenders are hot air outlets for the engine compartment. Owners: Herb Rothman and Ted Davidson.

French. I went to New York and met Dali in his hotel suite. That was an interesting meeting, seeing his room and how it was arranged. He had a coffee table made out of a slab of marble about seventeen inches high and about thirty inches long with irregular edges. Some holes were bored in it, with daisy-like flowers in the holes. It was really interesting.

I asked him what he would want if we used his drawings. He said all he would like would be the Cadillac of his choice. That was easy for us to do, so we agreed to that.

At the end of the meeting, he said that he would send us some drawings and I said I'd be glad to look them over. The drawings did arrive a few weeks later. There was a little note with them that said that when a Cadillac is born it's like a newborn babe, it's unclothed. He said the Cadillac is a regal car. His design reflected all this. From the beltline down, his design for a Cadillac had a purple robe with gold tassels all around it. It had two sets of window glass in it—one set would go up and down; and it had moons and stars on it. Dali said that, to keep it legal, this glass would disappear down when the car was just driving around town. But when you parked it or somebody was in the back of the car who wanted privacy, these windows could be put up. This would also give a Salvador Dali flavor to the car. The name of Dali's car was the Debutante.

We figured the car was not something we'd want to build. We wrote him a letter, thanked him, and sent the drawings back.

About a year later, Curtis had a Buick made for his daughter's coming-out party, or something like that, and he called it the Debutante. No sooner had the car been announced than we got a letter from Dali thanking us for recognizing the value of the car's name. He said he would like an Eldorado Brougham as payment. I don't know how they finally settled it, but I never saw him again. I believe he did get a Brougham.

Author: Do you think his offer to do a design was inspired by seeing the Eldorado Brougham at the auto show?
DA: I'm not sure, but I don't think so. I think he just felt that it would be good publicity for him and so forth, and it would if he had had something that was salable. Conceivably, Cadillac and Dali could have put something together, maybe for a show, that would have been interesting. Although, his work was tops in his field.

Author: Were you involved in that air suspension on the '57 and '58?
DA: A little bit, yes. The "praying mantis"? It was before its time It was a very nice riding car and it really did provide a fine ride.

Author: The other thing that really impressed me was that you could completely control the trunk lid from the driver's seat—both opening and closing it—and also you had those two memory seat positions, and that was almost forty years ago.

1958 Eldorado Brougham engine compartment. The engine compartment was so crowded with the basics and the numerous accessories that the battery had to be mounted in the trunk. The 1957 was powered by a 325hp version of the 365ci Cadillac V-8. For 1958, horsepower was increased to 335. Also, in 1958 the 1957's dual four-barrel carbs were replaced by three two-barrels. Owner: Garth Higgins.

DA: Yes, the memory seat was one of my contributions. I hold the patent on that. It's a dual patent. I hold the patent with Dave Campbell, a Fisher Body engineer. I actually drew the thing up and had to get somebody to make it and to put it together in a car. Fisher Body volunteered. And to the best of my knowledge, that's the only patent that General Motors went to bat on for somebody infringing.

Quite a few years later, Mercury came out with a memory seat and it was almost identical. They just quit building it.

Author: Well, I thought that was incredible for that to have been in a car so long ago.

DA: You see, there were no chips or anything like that. All there was was a whole bunch of wires. If it con-

1959 Eldorado Brougham concept proposal. Although the 1957 and 1958 Broughams were magnificent automobiles, GM lost money on every one they sold. Several concepts were considered to replace the series. The idea that finally prevailed was to design a special car for a production chassis and have it built in Italy. Some have said this was to save money on labor, but even if this idea had been considered, transportation of the chassis to Italy and the car back to the U.S. would have eaten up any savings. This is one of many renderings presented for consideration. The overall appearance of the design is appealing, but the abrupt and oversized tailfins seem inappropriate for the rest of the design.

1959 Eldorado Brougham. Hand built in Italy by PininFarina (the name was later changed to Pininfarina), these Broughams did not have the distinctive styling of the 1957 and 1958 versions. In fact, many of their design characteristics were picked up by later Cadillacs. Standard 1959 chassis were tested and specially fitted for the Brougham body work done in Turin, Italy. After their return to the U.S., finishing touches at Cadillac completed the manufacture. The Broughams did not have the huge nacelled fins of the other 1959 Cadillacs, were fitted with a large flat windshield, and were set off by a distinctive delicate roof. Only ninety-nine were built in 1959. One story has it that 100 were to have been built, but one chassis was dropped in the harbor during shipping. The Pininfarina "F" logo appeared on the Broughams of this year, but not in 1960.

nected when you'd have a hot wire someplace in an arc, this thing could move and it would hunt for it and it would stop at that place where there was a dead wire. It would stop at that dead wire where the thing was. So we had a bundle of wire going from the controller where you set the thing. And you had two positions. The mechanism would memorize for his or hers.

Author: Well, now on the trunk opening thing, was that done in your group?
DA: Yeah, we worked that out with Fisher Body.

Author: And why did you put the battery in the trunk?
DA: Well, there wasn't any other room for it. It was a bad place for it with the long cable and the voltage drop. Everything for cranking the car was, of course, up front, but we just didn't have space for it any other place.
Author: Your engineering group had to work out those aluminum bumpers for the '57 Eldorado Brougham?
DA: That was something, too, especially to get somebody to chrome-plate them. That was an innovation, that plating.

Author: I don't see how you did that. Those were intricate curves in the bumpers. It must have been difficult to get them done without breaking the metal.
DA: That was why we used aluminum, because of those intricate shapes. I can't remember who—I guess we worked with Alcoa on that—but it was an innovative thing, chrome-plating those things.

Author: And on the '59–60 Brougham, you said the rear window that went back part way for the door to open, that's funny because the only other car I know of that does that is the '61 through '67 Lincoln Continental convertible. I have a '61, and that window mechanism can be a problem. I just wonder if Ford got that idea from GM.
DA: I'm sure they bought one of those Broughams and went through it as quickly as they possibly could, because we'd do the same thing. I remember when we got their first Mark II. We got one of those and looked it over.

Author: Well, how does a body engineer look that over? Did you take anything apart or did you just look?

1959 Cadillac Eldorado Brougham. This detail shot shows how the unique rear quarter window moved several inches rearward automatically when the rear door was opened. In the 1957–58 Broughams, the windows in the rear doors would automatically partially lower when the rear door was opened. The tightness of window closure has always been a problem in four-door pillarless construction. When Lincoln made its 1961 four-door convertible, the rear window had to automatically descend a few inches when the rear door was opened, just to clear the seals between the window and the top.

1960 Cadillac Eldorado Brougham. Two views of the Brougham, not much changed from the previous year. Only 101 of these were built. Fewer of the 1959 and 1960 Broughams exist than of the previous series both because of the smaller production numbers and because of the innate fragility of the techniques necessitated to build such a small number of cars. The 1960 stock Cadillac grille, no hood ornament, lower fin, and the deletion of the Pininfarina insignia were the main points of difference between this model year and 1959.

Cadillac Eldorado Brougham "Jacqueline" proposal shown at the Salon de l'Automobile in 1961. The Pininfarina design firm saw the Brougham program drawing to a close, so it proposed two more Broughams, both labeled "Jacqueline" after Jackie Kennedy, to be built by the company beginning in 1961. One was the two-door version shown here and the other was a four-door. The design was not picked up by GM and the Brougham name was put on the shelf. A form of the Eldorado still lived, but the ultra-expensive Eldorado Brougham concept quickly disappeared with the refusal of this proposal.

DA: We took some things apart. In those days we used to get all of our competition and tear the cars right down to the nuts and bolts. We had a place as big as two football fields with long tables in it and there would be each manufacturer's car laid out piece by piece.

Author: That's incredible. Dave Holls said that the picture comparing the Mark II with the '57 Eldorado Brougham was the first picture that was done on the turntable out in the patio. He said that it was like a car from Mars parked beside a product of the Duncan Phyfe factory. But, after the Brougham program, unfortunately, it looked like Eldorado just became a special Cadillac.

DA: That's what it did. It was a convertible DeVille and was, like I say, the extruded aluminum or whatever we had on the lower body. Some of the latter ones had stainless steel moldings that were specific to it on the hood and body.

But the first front-wheel-drive Eldorado was distinctive, the 1967. We came out a year later with front-wheel drive than Oldsmobile did. We were ready to go out earlier, but we still didn't feel that we had everything quite in shape for a front-wheel-drive Cadillac for the '66 model year.

There was a lot of opposition. If you look at a front-wheel-drive car from a hot rodder's standpoint, when you accelerate, the load goes on the rear wheels where you want it so they won't slide. When you're stopping, the load goes on the front wheels and the rear wheels, and you get very light. It's hard to distribute the braking. So there were a lot of things that had to be worked out. We had to get it done right. I think the car that we finally came out with was a real nice car. It caught everybody's imagination. It was a nice looking automobile, too.

Author: What was Carlton Rasmussin's [then Cadillac

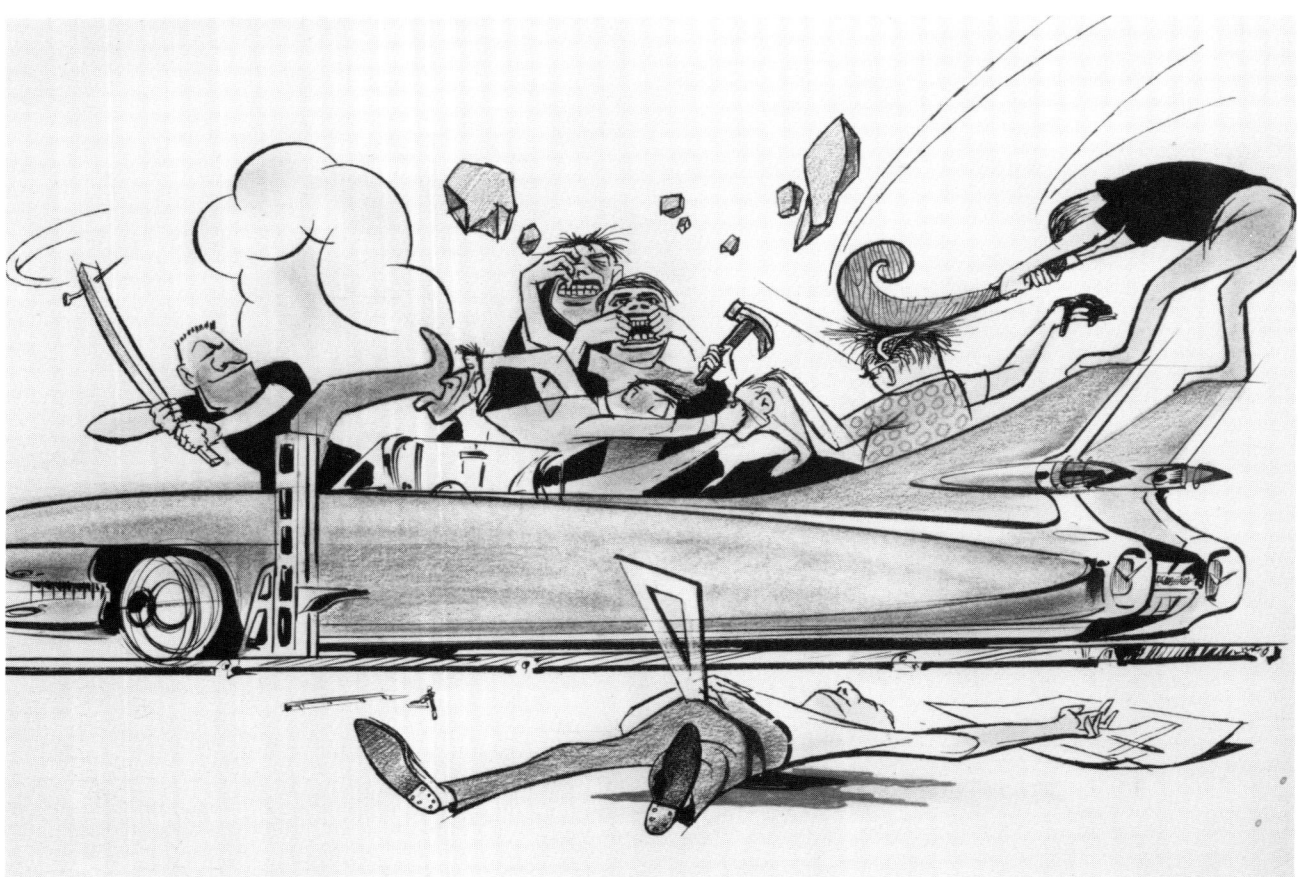

This drawing by one of the Cadillac Studio designers is a tongue-in-cheek look at a minor disagreement over the direction the design shown in the clay model is going. Major players involved in the 1959 Cadillac design, included Ed Glowacke, Dave Holls, Bob Scheelk (at that time responsible for the interiors), and, toward the end, Chuck Jordan.

Chief Engineer] primary contribution to the engineering of that car?

DA: He and I were both assistant chief engineers at that time. He had the chassis, the engine, and the front-wheel-drive mechanism—all that business. I had the body and the experimental and the forward planning. When Carl took over from Fred Arnold, he gave me the whole bit. For a year or two, I was the only assistant chief engineer and had the whole automobile until Bob Templin was finally brought into the picture.

Author: Did you have to make any corrections for the second production year, or did you make corrections all along?
DA: We didn't make any corrections after the start of production, but in the second production year we tweaked it a little bit. I think we put lights in the north of the fender. They were plain paddles sticking out there. And, we put parking lights/turn indicator lights in those fender noses and changed the molding a little bit. There wasn't much else done to it. The engine was being improved all the time and there wasn't much change during the second year. A model change poses a problem with service and so forth. But we had done a pretty good job, did our homework on it, and didn't have to make any drastic changes before the next model change.

Author: Since you had asked for an extra year, I assume you had more time than you would usually have in introducing a new car?
DA: We did, so we had it in good shape when it came out. We were told the chain, you know, the Morse Chain, would never work. This is part of the front-wheel-drive mechanism driven by a big wide chain. I have yet to hear one of those fail. That amazed a whole bunch of people.

Author: Now the '71 Eldorado—that's the bodystyle which was available in a convertible, you know—were there any problems with that one?

1959 Cadillac design underway. Here a designer is preparing sketches of various ideas for the design of the 1959 Cadillac. The top executives at GM at the time had laid down a rule that for the 1959 model year all full-size cars would share the same cowl, the most expensive part of the car to build, and many other basic body components. This would save a lot of money for the corporation. It would also pretty much dictate that the Eldorado would share the same chassis and body design as the other cars in Cadillac Division.

DA: It's a little larger. I think we widened the tread a little bit and we got the treads in line so that they were off of the front and rear the same. I don't mean that they were out of line, but they weren't exactly the same track. And, no, it was still very much a similar kind of a car. We built a lot of four-door sedans with that setup.

We'd go to the Greenbriar and build a couple of DeVilles with front-wheel drive in them. We'd drive around and no one knew that it was a front-wheel-drive car. We said that this is a thing of the future, we could put front-wheel drive in the full-size Cadillac. We told the executives that if they were afraid of that, we could put it in both

55

ways. It would cost so little to do it because all we had to do is an underbody.

The thing that finally killed that idea was the transmission. They didn't have capacity to build enough transmissions. They would have had to make the Hydra-Matic plant much larger. That's the final reason that we didn't do it.

But we built some real nice cars. Styling built a thing they called the *Black Snake*. They lowered it, and everything throughout the body was just as smooth as could be and with a front-wheel drive! We'd drive around with Charlie Cheyne. He was an advocate of the front-wheel drive. I still have my pass to the proving grounds in the Tech Center that Charlie Cheyne signed. Cheyne was a vice president in charge of the Tech Center. He came from Buick, I think. He was a very colorful person.

Author: But the *Black Snake*, was that an Eldorado or a Coupe DeVille or...?

DA: It was a four-door sedan made from a Sedan DeVille. When we came out with the first small car (the '75 Seville), we went all over the world to find somebody to build a front-wheel drive. That was in the early seventies. We went to Australia [Holden], to Germany [Opel], and all those people said their front-wheel drive would never go. No way. And, finally we had to settle for domestic components.

George Elges [Cadillac general manager] drove our prototype front-wheel drive version and he liked it. I re-

1959 Cadillac clay model. A December 1956 design study for the rear of the car. Actually, this is two studies because the right is different from the left. Note that the tailfins are much like those of the 1957 - 1958 Eldorado, being inboard a bit and of roughly the same shape.

1959 Cadillac clay model. Even if the profile of the car had been pretty much settled, there seems to have been some question as to what to do with the car's rear. This photo was taken in March 1957 and shows two rear end possibilities on one model.

1959 Cadillac clay model. By February 1957, when this model was prepared, variations in the chrome accent designs for the Eldorado were being tried. The overall profile of the car seemed to be close to what it would be at production time, but the shape of the tailfins and the chrome and medallion placement would change. Four-door models are shown in these photos because this was the basic body design that was developed before working out particular models such as the Eldorado Biarritz and Seville.

member going down and picking him up at lunch at the General Motors building and driving around in it. But we ended up keeping the conventional drive on the car.

It was interesting that when we came out with the car—we were shooting at Mercedes, of course—we sold more in California alone than all of Mercedes [sales] put together that year.

Author: The '78 Eldorado was the last of the '71 bodystyle. And, the '76 was the last year they made a convertible.

DA: I was very active in getting that convertible into production. I had to go find parts and everything, then scrounge them so that we could continue to build a convertible. I did this because the government had already reversed their stand on convertibles. But I knew convertibles were good cars. I always wanted a convertible. I had a number of them myself. But that front-wheel drive job was the greatest!.

Author: The '79 was still front-wheel drive. Was that the last Eldorado you worked on?

DA: Yes. We had plans. The last two years (to July 1978) I was director of all forward planning for Cadillac.

Author: But at least you didn't have to undergo the indignity of the '86. So many people I've talked to are really, you know, bitter about that '86.

DA: You're talking about the downsized Cadillacs, beginning with the Cimarron which was earlier than '86? I have one of those, too. I almost have 80,000 miles on it and I've never had any trouble with it.

Author: I'm sure you know the problem everybody was talking about.

DA: It was a mistake to drastically downsize the full-size Cadillac to that extent. Yet, many liked it. I was really against even doing the Cimarron. I was against using the Chevrolet body as a basis when we couldn't get Opel in Germany or Holden in Australia to do it. The 1975-1/2 Seville was a reduced-size Cadillac. But it worked out fine. I mean, it did a fine job. It was well-engineered and everything, but it was still not a Cadillac for me. I think that we could have done a lot more.

Author: When you were head of forward planning, what kinds of things were you looking for as far as Eldorado was concerned?

DA: We were certainly sure we wanted front-wheel drive. We were working with what could be done with four-wheel drive. I believe Cadillac built one or two cars with four-wheel drive. They were also looking at four-wheel steering. They investigated all these things that you see the European cars have done.

And we had the wherewithal. Triad and American Sunroof were among the engineering companies that built experimental stuff, and I worked with them. I even spent time with them a couple of years after I retired [in an advisory capacity] for Cadillac. They came up with many helpful things.

But—and I emphasize that this is strictly my opin-

1959 Cadillac clay model. The tailfin nacelles (the bullet-shaped structures that held the taillights on the fins of the 1959 Cadillacs) of this model were photographed in March 1957 and are part of this design, along with a less elaborate side air scoop. The front end is approaching the form seen at production time.

1959 Cadillac clay model. Taken in September 1957, this photo shows work on a demonstration of two more proposals for the car's rear end. Note the pointed lenses on one side and the flatter ones on the other.

ion—I have to say that I don't agree with all this emphasis on wind-tunnel testing. For the average car, modifying design just so it can go fast on the freeway and save fuel, it doesn't add up. All of the fuel that the majority of cars burn in great amounts is done in traffic conditions and in the city—35 and 40mph, stop and go. The wind tunnel doesn't have a damn thing to do with that.

All the car companies lost a lot of marque individuality by pushing this grade point zero, this air slip thing. But, wind-tunnel testing is required for noise reduction.

Author: Every time I see one of these air foils on the trunk lid of somebody's car, I wonder at what speed something like that really has a cost-effective effect—seems like I read somewhere there's no real effect until about 115mph.
DA: On some European cars, the air foil doesn't pop up until you get going around 100mph. On some, it even obstructs the motorist's rearview vision.

Author: When we spoke earlier, you mentioned there was a story behind your memory seat invention.
DA: I flew B-25s. I've got 1,500 hours in them. There's another story there. What I did, I was an instructor pilot. I never went overseas. I was in Pampa, Texas, which is up in the panhandle of Texas. We couldn't keep B-25s flying because the mechanics didn't know how best to maintain them properly. We got hold of the commanding general there and told him we worked out a similar problem in the automobile business. He said, "You've got the job. You work out whatever it takes to get these guys trained."

So they gave us a bakery, wiped out the bakery machinery, and made it into a machine shop. We made training aids. We cut-away engines. We made operating cutaway propellers. In other words, we set up a school and we got results right away. The general was so impressed with it that he told us to go into production and make training aids so there would be enough sets of instructional aids for all the places in the United States where they were servicing the B-25 planes.

Author: Yet another factory?
DA: I had a B-25 assigned to me and I had returnees, master sergeants, and specialists that were experienced machinists. I had experienced electricians and sheetmetal and plastics technicians. I had the cream of the crop.

Author: So, did you go back to teaching after you set that up?
DA: No. That kept me busy. I did the designing on how these teaching aids would be built. We did cut-

1959 Cadillac full-size drawings. These photos show full-size renderings of the car done in October 1957 and later when the design for the production car was being tried with various chrome side treatments and the proportions evaluated in full-scale.

aways, made them remotely controlled. Again, we didn't have any little things like transistors and computer chips like we do today. We did it all with what was available. That's where I got the idea for the memory seats we used on the '57 Eldorado. We had a model of a plane, really a silhouette of an airplane about three feet long. You could lift the tail up, for instance, and a bunch of hidden cables would move the dials to show the increased rpm that you get by diving the plane. The little electrical circuit would turn the motor that turned the blades on the propeller to make a bigger bite as it dove—a student could see how all these things worked together. You weren't sitting in an airplane [but you could see how it worked]. Designing the memory seat mechanism, when I got out of the service, was just another step beyond that, and you could see how it worked—and it did.

March 1957 clay model of a design idea for a 1959 convertible. This would have been an early stage toward developing the Eldorado Biarritz's design.

October 1957 clay model of a two-door 1959 coupe, an early stage toward developing the Eldorado Seville's design.

1959 Cadillac Eldorado Biarritz production convertible. With all the elements of the design worked out and produced in the production materials, the car looks much more sleek and elegant than the same form in the clay models. Note that as in 1954, the Eldorado is again based on the regular Cadillacs, except for some special chrome, an upgraded interior, and a few accessories.

October 1957 clay model of a two-door 1959 coupe, an early stage toward developing the Eldorado Seville's design.

Chuck Jordan joined Cadillac Studio late in the 1959–60 program. Bill Mitchell had told him that if he really wanted to get anywhere as a designer he had better leave truck design and start designing cars. Ed Glowacke (Cadillac Studio chief designer from June 1951 to August 1957), Dave Holls (the major force in this design), Bob Scheelk (who did the interior), and others on the Cadillac Studio staff had wrapped up the 1959 design and had begun the 1960 when Jordan came on board as the new Studio chief in October 1957.

1959 Cadillac Eldorado Seville production coupe. This is the hardtop version of the 1959 Eldorado. The distinctive Eldorado side molding dramatically swept from the front of the door to the rear of the car, top and bottom, and was used on both the Biarritz and Seville.

Only from the great traditions of Cadillac could there come a motor car as surpassingly fine as that portrayed below —the 1959 Eldorado Biarritz. Luxurious beyond description, it offers spectacular performance and handling ease. Every known motoring advancement makes each journey memorable. Standard equipment includes a custom engineered 345-horsepower engine, air suspension, electrically powered front seat adjustment, electric door locks and window regulators, power steering and braking, radio and heater. Interiors are offered in deep-grained Cardiff and Florentine leathers in tones of bronze metallic... blue metallic... gray metallic... slate green metallic... black... white... and red.

THE NEW STANDARD OF THE WORLD IN SPLENDOR!

Eldorado Biarritz

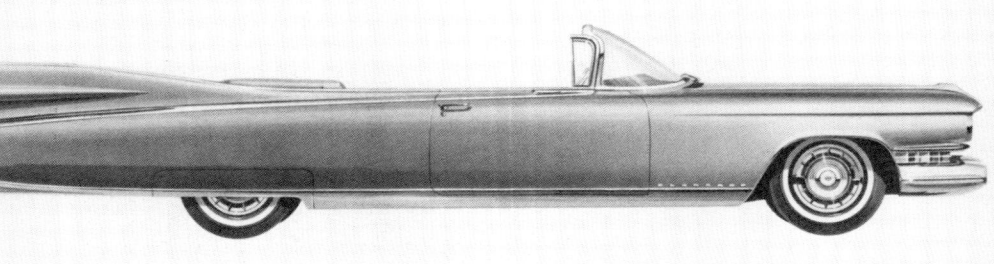

As if the cars were not long enough to begin with, these renderings of the production models appear even more elongated. Despite the radical design, all sales material such as this part of a special brochure was done with great taste and elegance.

Full-size drawing of the 1959 Cadillac.

Irv Rybicki Interview
Cadillac LeMans Show Car and the 1979 and 1986 Eldorados

Irv Rybicki is the former vice president of design at GM. He retired in 1986. Rybicki studied at a couple of specialized art schools in Detroit before joining the corporation in 1945. He had worked in several divisions besides Cadillac, but was involved in the design of the Cadillac LeMans show car, the 1979 Eldorado and, reluctantly, in the design of the downsized 1986 Eldorado. Rybicki was responsible for the first facelift of that car.

Author: You worked on the Cadillac LeMans show car?
IR: Yes. I worked on the exterior some, but I did the interior on my own. Harley Earl worked on the exterior of the car closely with four or five designers. Bill Mitchell was the chief. We talked it out and worked it out and Earl came in and looked at our drawings.

I remember one incident with Earl back then. We had the rear deck very low and there wasn't a hell of a lot of room to stack a top. I kept putting the top down and I said, "Gee, Mr. Earl, we're not going to be able to get the top in there."

Earl said, "Get Freddy Walters in here. We'll ask him." He was the guy who was in charge of engineering at Design Staff.

Freddy came in and did a lot of scaling. "Mr. Earl, if you move that down a sixteenth of an inch more, we're outa gas," Freddy said.

Earl said, "Thanks, Freddy." But after he left, Earl said, "Move it down an inch!"

We moved it down an inch and Freddy got the top in. That's a true story because I was right there when it happened.

As I said, I did the design for the interior of that car. I had done an interior sketch with my name on it, and hung it on the wall in the studio. Jules Andrade, Earl's right-hand man, saw the sketch one day and asked me to take it down. He said that he didn't like it, but that Earl would if he saw it. I told him I'd take it down after he left. When Andrade left the studio, I just left it up there. Sure enough, Earl came in that afternoon, saw it, and really liked it. He said, "Guys, do this one full size."

So Jules was right about one thing, Earl liked it.

Author: What was the first Cadillac Eldorado that you worked on?
IR: I worked on a couple at different times in my work on the complete Cadillac line, but I don't remember the specifics.

I do remember working on the Eldorado that became the '79. Jack Humbert was my assistant—and there wasn't a better one in the business anywhere. He died of leukemia about a year before I retired.

Irv Rybicki.

When Jack and I first saw the car, Wayne Kady had done about 30 percent of it. We told him to put a cover over it and forget it, that we were going to start over.

The studio started over. Jack and I had been talking about it in the office and we had a pretty good idea of where we wanted to go. So we laid our design out in the studio, cut the templates, and started modeling. It was looking good. Kady didn't like what we were doing until a couple of months had passed.

Author: That takes us up to the downsized Eldorados and your fights over that?
IR: I hated that. I even hate to think about it. I was vice president for Design at General Motors at that time. The arguments I had with the general manager of Cadillac and his chief engineer were out of sight. We just didn't get along. They kept telling me the downsized car was a Cadillac and I told them it was nothing. And they asked, "Irv, what in the hell do you know about the sales of an automobile?" I said, "I think I know what Americans buy, and they won't buy that." And they laughed at me.

Author: So, as I understand it, the whole force behind the downsizing was from Cadillac Division? Design Staff didn't want any part of it?
IR: Not that small, no way. They were scratching around at 190 inches of length—at the time the Chevrolet Caprice was 215 inches in length. Now that was an old Chevy and we're talking about a new Cadillac, I understand that fully. But they're both going to be on the street at the same time.

But when you package a car at 190 inches and you look at the kneeroom, and you look at the volume in the trunk, you haven't got a Cadillac. You haven't even got a Chevrolet. I don't know what the hell you've got!

This guy who was the chief engineer, Bob Templin, he was a big joker. He'd come in and tell you a lot of stories and jokes, but make no comment about the car and

1959 Cadillac Eldorado Seville production coupe. This is the hardtop version of the 1959 Eldorado. The distinctive Eldorado side molding dramatically swept from the front of the door to the rear of the car, top and bottom, and was used on both the Biarritz and Seville. The big design feature for the year was, of course, the fins. What most people don't realize about these cars is that they were wonderful to drive and handled well. Also, when driving a 1959, one had excellent visibility and could see all four corners of the automobile.

where we were going—or what we were doing. He'd just sort of stand in the background. The general manager would have to do the fighting.

Well, when we proved to them we couldn't do it at 190, they went up to 196.

I refused to release it. It was still too small. They came in on release day and asked if I would release it. I said, "Hell, no, that's not a Cadillac. My job's on the line if I do that. Forget it." The next day they marched in with the president of GM, Jim McDonald. I greeted all of them and Jim looked over the car they wanted released. But I had another one sitting there that was 209 inches long, which was considerably shorter than the Cadillac we were going to outdate. I didn't show the 209 inch car, I had it covered.

Jim McDonald had looked at the 196 inch car and he said, "Let's go in the office and talk."

The two of us went into the office. He sat at my desk. I sat in a chair.

He said, "You've got to remember something, young man."

I said, "What's that?"

He said, "For the general manager, Cadillac is *his division*. You've got to do what *they* want."

I said, "Even if it's as asinine as that thing in the studio?"

"You may think it's asinine. They don't. So release it. Would you please?"

I said, "Jim, I oughta make you sign a document that you've just asked me to release that thing."

He said, "Well, I wouldn't sign it, but I want you to release it."

I released it. They produced the car, and it didn't go anywhere. It was a loser.

Author: But you were kind of stuck with it?
IR: We were stuck with it and couldn't budge.

But after a few weeks, I got to thinking, I know the chairman of GM, why don't I call him and give him the pitch? So I called Roger [Smith] and told him I had something very important I wanted to discuss with him in the Cadillac Studio. I told him that if he could find some time anytime over the next week or two, I'd appreciate it. It got quiet at the other end. He said, "I'll be out at two o'clock tomorrow afternoon."

He came out. I had the 209 inch car I wanted uncovered and we had a buck to represent the interior and the kind of leg room that would be provided and the gain in the rear compartment and the volume in the trunk.

I took him through the whole pitch. I showed him the clay that Cadillac wanted and the clay we were pushing. He looked at both.

"Gee, Irv, there's no comparison," he said.

I said, "Would you tell that to Cadillac?"

He said, "I'm going to ask those guys to come down here and release that car for the next go-around." Which he did. He called the general manager of Cadillac. He told him to get his engineering people in there and get going.

The Cadillac people came in the following day and

1960 Cadillac Eldorado Biarritz convertible. For this year, the fins were reduced and their nacelles removed. Both the front and rear ends of the car were changed slightly. The overall appearance was smoother than the previous year. The Eldorado accent side molding was reduced to a dual, fine chrome line.

nobody would talk to me. They hated my guts. I had gone over their heads and now they were thinking they were going to *have* to do this thing.

When the 209 inch car got out on the road, it didn't fix our sales volumes entirely, because once you lose the confidence of a customer, it takes a few years to get him back. But it did turn things around some. We were selling more cars than we had been with that little thing.

Author: So the 209 inch car was the "second generation" of the downsized Cadillacs?

IR: Yes. It established the next generation of downsized GM cars and trucks.

Author: What do you think is the future of Eldorado?

IR: I think the Eldorado and the Riviera will hang in there. The Toronado will be gone in a matter of months. They're replacing that with the Aurora. The last car I owned when I was still at GM was a black Cadillac Eldorado with gold striping. It was a great car. I loved it.

1960 Cadillac Eldorado Seville coupe. From this view, the Seville looks to be a huge and much more tame Eldorado than one would have come to expect in, say, 1957 or 1958. Certainly, it wasn't as easy to distinguish from other two-door Cadillacs of the same year.

AMERICAN ♦ CLASSICS

Chapter 4

1961–1970: A Legend Reborn

A Time of Badge Engineering

The beginning of the sixties was a low point for the Eldorado name. The Italian-bodied Eldorado Brougham ceased production in Italy during the 1960 model year. Pininfarina had made a proposal to GM for the 1961 production run of a new Brougham called the "Jacqueline," obviously in honor of Mrs. Kennedy. Although these pillarless, finless Cadillacs were remarkable for their time, Cadillac decided to pass on the proposal.

The Eldorado Biarritz and Eldorado Seville had existed only as a badge, chrome accents, and interior packages for the 1959 and 1960 model run. The 1961 model year was not much better for the Eldorado name; in fact, it probably could be termed a step backward for the concept. A major difference was the sad fact that the Eldorado was available only as the Biarritz convertible in 1961. The Eldorado was just a part of the designated Series 63. Regrettably, the badge engineering package program was all that existed under the Eldorado name right up through the 1966 model year. All this was not from lack of interest in restoring Eldorado to its previous glory. Designers presented proposals for possible all-new individual Eldorados, but none was carried to production.

As mentioned earlier, Chuck Jordan became chief designer at Cadillac Studio when the 1959 Cadillac was being finished. Bill Mitchell was determined to reduce the amount of chrome on GM's cars and Chuck Jordan, as chief designer, joined him in this new philosophy. As he saw it, Jordan's aim was to restore Cadillac's stylistic reputation. He insisted that the 1960 model be a toned down version of the flashy 1959, and began to lay the groundwork for what he considered a line of elegant, but sporty, Cadillacs. With his assistant, Stanley Parker, he set out to accomplish the first step with the development of the 1961 Cadillac.

The 1961 Cadillac and its closely related 1962 facelift version are unique cars because they have both upper and lower tailfins that rise, almost organically, from their rear fenders. Much reduced from the huge 1959 version, these fins accentuated the cars' lines, giving a flow and thrust not found on any other car. The upper tailfins on the 1962 Cadillacs were slightly lower than those of the previous year.

The lower fins, called "skegs" by the designers, first appeared on the 1959 Cyclone show car and then the Italian-built 1959 Eldorado Broughams. In fact, the 1961 and 1962

The 1958 Firebird III, a show car and a technical laboratory on wheels. Try parallel parking with this beauty! Many critics saw this—and its two predecessors, Firebirds I and II—as deadends in the evolution of design, but in reality they were free flights of the imagination. To describe the practical technical innovations incorporated in this car would take several pages, but few who looked at it could get past the jet-plane appearance. Even when you see the vehicle today, it looks like something from another world. Indeed, Harley Earl thought of it as the perfect car to drive to the astroport as the first leg of a journey to the moon. As you can see, there are several fins on the car, including what are called "skeg fins" dipping down to the front of the rear wheels. Similar fins would appear on the 1959 Cyclone show car, Italian-made Broughams, and the 1961 and 1962 Cadillacs. Notice the fin on the front fender that appears to arise out of the leading edge. Front fins never made it to Cadillac production cars, but the line arising from the leading edge of the fender did. (The photograph was taken beside the twenty-eight-acre lake at architect Eero Saarinen's $125 million GM Tech Center, now renamed North American Operations (NAO) Tech Center. Saarinen had inherited the commission to design the Tech Center from his father and did a masterful job of creating several splendid locations for automotive photography on the 900-acre site.)

Cadillac bodies had much the same overall form as the Italian-made Eldorado Broughams, the most immediately recognizable difference being in the front end.

Originally, the designers tried versions of the skegs that were themselves going to be much larger than their final production form. But Jordan was looking for something elegant. He wanted Cadillac to be sporty, yet regain some dignity in design.

For 1961, the Eldorado Biarritz convertible was similar to the $1,000 cheaper Series 62 convertible. Two ways to identify the Eldorado were the "Biarritz" nameplates on the

1961 Cadillac clay model. By December 1958, plans for the 1961 Cadillac had reached this point. Fins were greatly reduced and the car had taken on a more compact appearance. Bill Mitchell had told Cadillac Studio Chief, Chuck Jordan, to tone done the fins and return Cadillac to its lofty position in the world of automotive design. With his assistant, Stan Parker, Jordan tried to create a car that was "elegant, but sporty." The wraparound windshield was now a thing of the past and GM had a mandate for extensive body part sharing amongst the various divisions. The lower body "skeg fin" and the top fin, which arises out of the leading edge of the front fender, tie this car's design together. The flow of the crisp lines, the four fins, and the dagger-blade rear fenders of this model make it appear to be moving even though the form is sitting in a heavy clay buck.

1961 Cadillac clay model. By early 1959, the designers were trying the fins back further on the rear fenders and smaller sail panels, which resulted in a larger rear window. The line from the leading edge of the front fender to the top rear fin is now reduced, which, together with the change in fin position, has a negative effect on the thrust of the design.

front fenders just behind the headlights, and the special leather interior of ostrich-grain hides with premium Florentine leather trim. The 1962 model sported a narrow chrome side strip that swept from the top line of the door to the rear of the car. Bucket seats were a no-cost option along with a list of standard luxury features and, of course, a unique Eldorado interior completed the package.

Production for the 1961 and 1962 model years was identical at 1,450 copies each, a mere 165 more cars than were produced in 1960.

A dual braking system was introduced in 1962, as well as cornering lights. These features and the three-phase rear lights make the 1962 Eldorado a preferred early-sixties convertible for collectors today.

By the 1963 model year, Chuck Jordan and his design crew were well on their way to reaching the goal of a simple, elegant Cadillac. The fins were lowered even more, the skegs were deleted, and the overall design was smoother, especially the clean sides which were almost understated. Only the front of the car reminded one of the 1959 era.

There was no special chrome on this year's edition, only block letters spelling out "Eldorado" on the fenders just behind the front wheels and a wreathed Cadillac crest on the sides near the rear bumper. In fact, the Eldorado had less side chrome than the Sixty-Two convertible, which had a narrow horizontal piece of chrome running from above the front wheel all the way to the rear bumper.

The big news for all 1963 Cadillacs was that the old V-8 that had been introduced way back in 1949 had finally been replaced by a smaller, lighter, redesigned V-8 with the same horsepower and displacement as the engine used in 1962.

On the surface, the 1964 was a minor facelift of the 1963. But this car, the last year of *true* Cadillac tailfins, came with a host of changes and refinements. Instantly recognizable from the front because of a wide horizontal divider, painted the same color as the car and crossing the grille in the middle, all 1964s in the Cadillac line were crisp and fresh. Cadillac Motor Division said the 1964 was "more than a hundred ways new." The V-8 introduced in the 1963 model was upgraded to 340bhp with a displacement of 429ci. This

1961 Cadillac Eldorado Biarritz production car. This is one of the most surprising and satisfying Cadillacs to see in the flesh. The simplicity of line and the skeg fins really make the car's design. Besides the more luxurious interior, there were few design cues to distinguish the Eldorado from other Caddy convertibles. The glob of chrome right behind the headlight on the side of the car spelled out "Biarritz," but did nothing to enhance the beauty of an otherwise striking design. The Eldorado was, once again, available only as a Biarritz convertible.

1961 Cadillac Eldorado Biarritz, production car. By the time the design reached production, much of the flow of line seen on the early clay model had been restored. The upper line on the side of the body that arises from the front of the fender no longer flows into the top edge of the fin as it did in the model, but now is well below the fin and sweeps to the rear of the car just above the bumper's taillights. From this angle, the similarity between the 1961 Cadillac and the Italian-made Broughams is quite apparent.

1962 Cadillac Eldorado Biarritz production car. For the most part, this was a facelift year that cleaned up a couple of things in the original design. The "Biarritz" chrome name glob was gone from the front fender. The line that emerged from the leading edge of the fender now was partially highlighted by an accent line beginning in an Eldorado-crest shield near the front edge of the door and sweeping back to the end of the car. The rear of the car is more homogenous than the 1961. The result was one of Eldorado's most powerful designs, every component of the design worked. The dual braking system, cornering lights, and premium Florentine leather trim made this an attractive buy when it was introduced, as well as for today's collector. Only $1,000 more than the Series 62 in this model year, an identical number (1,450) of Eldorados were produced for 1961 and 1962. Here the light emphasizes the crisp lines that seem to make the car surge forward, even when standing still. The skeg fin is an integral part of this, one of Eldorado's most successful designs.

Left, 1962 Cadillac Eldorado Biarritz. One of Eldorado's most powerful designs, every component of the design worked. The dual braking system, cornering lights, and premium Florentine leather trim made this an attractive buy when it was introduced, as well as for today's collector. Only $1,000 more than the Series 62 in this model year, an identical number (1,450) of Eldorados were produced for 1961 and 1962. Owner: Mike Porto.

allowed some 1964 Cadillacs to accelerate from zero to 60mph in less than ten seconds. One automobile magazine had clocked a Coupe DeVille at 123mph in a test run. This was a banner year for looks and performance.

Eldorado also shined for this model year, regaining some of its distinctiveness. The name Biarritz was dropped after nine years. The car was now known as the Fleetwood Eldorado. Again, as in 1963, the car had no special chrome except for the block letters and emblems which remained in about the same location. However, the Fleetwood Eldorado was the only model to come without rear fender skirts. Instead, the chromed full rear wheelwell openings emphasized the sporty heart of the car. This open rear wheel treatment was seen nowhere else in Cadillac that year, except on the Cadillac Florentine show car exhibited in the General Motors Futurama Pavilion at Flushing Meadows, New York.

Opposite page, if Fisher Body had accepted the INFORA Top mechanism proposed by the engineers in Design Staff, the seats would have been wider. But this is still a luxury interior by anyone's measure.

1962 Cadillac Eldorado Biarritz, detail of production car door and end of rear fender. Here is a good example of detailing in the best of Cadillac design. Others might have just run an accent line of rolled chrome across the top of the highest crease on the side of the car. Cadillac mounted an elegant Eldorado badge on the door above the crease and then ran a double line of chrome with a painted center back to the end of the car, making a sharp downward dip to follow the line of the rear bumper. Owner: Mike Porto.

1963 Cadillac Eldorado Town Car, artist's design rendering showing an interesting roof treatment. Keeping many of Cadillac's design cues, the artist was able to experiment with an almost-fastback roof line and sporty, but sophisticated, wheelwells.

1963 Cadillac Eldorado, two clay models flanking a fiberglass model. The skegs have been deleted and the upper fins lowered. Elegant and sporty had progressed to smooth sides and a cleaner appearance. The front end still carried the character of the 1959, but in a quieter way. Chuck Jordan is standing by the rear tire of the clay model on the right.

1963 Cadillac Eldorado Biarritz convertible production car. Still no doubt about it, this was a Cadillac! This was the last year (for a while) that Eldorado would be linked with the "Biarritz" name. Cadillac produced 378 more Eldorados this year than for 1962. There were no side moldings, and the Eldorado name was spelled out in block letters on the lower part of the front fenders. A wreathed Cadillac crest marked the end of the rear fender. Wide chrome rocker panels, shared with the Sixty Special, emphasized the car's 223in length. Under the hood was the first all-new Cadillac V-8 since 1949, 50lb lighter than the previous year, but with a greater power-to-weight ratio.

When Cadillac introduced the 1965 models, no statement was made as to whether or not the tailfins had been deleted. That was left to the automotive press to decide. However, if you look at the evolution of the fins, say, from 1959 through 1965, they follow a steady progression of becoming smaller and less defined until, finally, for 1965 they are essentially the rear fenders themselves, and have no separate form of their own.

Designer Wayne Kady says that one of the first things he did when he went to work under the direction of Stan Parker in Cadillac Studio was to remove the fins. They were gone, but the car had reached the plateau of elegance that Mitchell and Jordan had foreseen way back when they passed judgment on the excesses of the late fifties.

The design of the 1965 is a masterwork of simplicity. Beveled sides accentuate the long, low character of these luxury cars. The aggressive front fender housings of the vertical dual headlamps enhance the simple elegance of the horizontal theme of the front grille, which itself is a jewel of fine lines and symmetry. The rear-end view has this same feeling of depth and beauty. The 1965 could be considered Cadillac's first step in the production use of what would eventually become "blade" fenders. If ever a design cried out the name "Cadillac," this one does. This design was no easy achievement because all the Cadillac cues remained, but the fins were gone. The car's design was a rolling contradiction that worked!

The Fleetwood Eldorado convertible was one-half inch longer than the previous year. The word "Fleetwood" appeared in block letters behind the front wheels, and there was no distinctive side chrome except for wide chrome moldings on the rocker panels which extended to the base of the rear fenders. The fender skirts returned for this model year. Of course, special Eldorado interiors were offered, this time in eight luxurious perforated-leather choices.

For 1965, Cadillac used a new perimeter box-section frame which allowed for a lower center of gravity. Related to this, Cadillac was able to use a straight driveshaft and new sound-deadening body mounts. Bringing all these components together resulted in a very quiet automobile on a rigid chassis.

For 1966, the Fleetwood Eldorado remained pretty

The Cadillac Florentine Two-Door Coupe was a modified production show car presented at the 1964/1965 New York World's Fair; it was the first Cadillac show car exhibited since 1959. Basically a customized two-door hardtop, the rear quarter windows retracted into the sail panel of the vinyl suede-covered formal top. Other special features included chromed wire wheels, open rear-wheel wells, handleless doors, high-back bucket seats covered in embroidered leather, a reduction of side body trim, and glass-hooded headlights.

Above and right, 1965 Cadillac Eldorado Convertible fiberglass model. This is a closing stage of the design process for the year's model. The photo on the right is the interior design buck for that year.

much in the same form, except that front bucket seats were offered as a no-cost option and all cars came with triple-stripe whitewall tires.

Without any fanfare at the time, this was to be the last Eldorado convertible until 1971. It was also the last rear-wheel-drive Eldorado for the foreseeable future.

A Fun Experiment with a Lasting Effect (XP-727 and Related Programs)

As mentioned, some proposals had been made from time to time in the early sixties for a uniquely bodied Eldorado to recapture this model's rich heritage. Chuck Jordan was in charge of Cadillac Studio and he wanted to come up with something that was truly different, sporty, and elegant.

As a fun project, he decided to look into the idea of a twelve- or sixteen-cylinder special-bodied automobile that would be reminiscent of the sixteen-cylinder Cadillacs of the thirties. Several designers worked on this project, many sketches were produced, and even a few models were built and photographed. Jordan emphasizes that this was a fun project that allowed the designers to stretch their imaginations and explore new realms.

Driving to work each day, Jordan dreamed of somehow building one of these behemoths to cruise around. He even considered things he knew would be "childish", such as somehow joining two eight-cylinder engines together in a row to make a sixteen-cylinder engine. As he says, this was a fun project.

In the end, a few sketches and models illustrated a design philosophy centered around a two-seater with an extremely long hood–almost "a selfish king car," as Jordan called some of the designs.

Despite it's fantastic nature, this project was not a waste of time, by any measure. Important design themes were developed and explored. An entire area of design exploration was made available for future Cadillacs. Indeed, one can see many things in these project renderings that appeared in production years later under the Cadillac crest, and, especially, under the Eldorado name.

continued on page 80

1965 Cadillac Fleetwood Eldorado convertible production car. America was stunned when suddenly a Cadillac appeared with no fins! Wayne Kady said that the first thing he did when he was put to work on the 1965 program was to remove the fins. But fins were a matter of definition, the protrusions and trailing blades of the rear fenders were still there, having followed a natural evolution of change from model year to model year. Since 1959, the fins had gradually shrunk and were now completely incorporated in the rear fender, which was in 1965 just a single line from the headlight to taillight with only one slight jump just before the rear wheel. Actually, everything about this car, except the basic engine which was only two years old, was totally new.

Not since 1948, then 1957, had Cadillac made such an across-the-board slate of changes. Most basic of all was the switch from the X-frame to a more rigid box-section perimeter frame. Cadillac announced that these were the quietest cars ever. Eldorado was one of the few models that came equipped with the new automatic leveling device as standard equipment. Having been given back its fender skirts, the Eldorado still looked much longer than the previous year, but had actually increased only 1/2in in length. (During this same model year, the Brougham name was resurrected to designate a formal roof package for the Sixty Special, quite a come down from what the name had meant in 1957 and 1958.)

The Massive Creative Process Preceding the First Front-Wheel Drive Eldorado

Throughout the model years from 1959 until 1966, there was no completely separate model labeled Eldorado. The cars bearing this name were only more luxurious models of the Cadillac line. Several proposals were presented for a unique Eldorado by designers from Cadillac Studio, but none were accepted. Chuck Jordan was chief designer at Cadillac Studio at this time and was working under Bill Mitchell's (who had succeeded Harley Earl as vice-president Design) edict to return Cadillac to its former design status. "Elegant and sporty" were the words driving the project.

Jordan later became executive designer (Bill Mitchell's first assistant in the chain of command) replacing Ed Glowacke who passed away in 1961. Jordan conceived the idea of designing a personal luxury car embodying such qualities that those seeing it would call it a "King Car." To drive such a behemoth, Jordan envisioned a twelve- or sixteen-cylinder engine, reminiscent of those massive powerplants Cadillac had built in the thirties. A twelve-cylinder engine was actually built and is in the Cadillac Museum in Detroit. Jordan dreamed, childishly he says, of somehow bolting two eight-cylinder engines together to make a sixteen-cylinder engine. Such a powerplant would require a long hood. He set some of the best General Motors designers to work on this project.

Over the years that followed, many designers were in and out of the project, including Stan Parker, Wayne Kady, and Jerry Brockstein. Jordan and Mitchell oversaw the operation. Some good ideas came from it. Eventually, an offshoot of the effort produced the design for the first front-wheel drive Eldorado, the 1967, one of the most beautiful and unique automobiles ever produced by Cadillac.

Above and right, May 1961 thru May, 1962 Concept Eldorado. One concept tried several front-end designs with a design that

had a sail panel sweeping back along the top of the rear fender to the end of the car.

Left, May 1960 Concept Eldorado (XP-727). This massive project involved several experimental designs (designated by various "XP" numbers). This early clay model incorporated some of the flavor of the 1957 Eldorado Brougham, but also included "skeg" fins found on the production 1961–1962 Cadillacs. If you ignore the skeg fin, the rear fenders are similar to the production 1963 Chevrolet. This model had the look of a sporty personal luxury car and was compared to the current Ford Thunderbird on the GM Tech Center patio.

Opposite page right, "V-12" is almost hidden here on the grille of this car which seems to be on the attack. This is probably the most extreme example of "blade fenders" seen on any models or drawings, but once you've seen these, you can understand the design concept that was incorporated into Cadillac and Eldorado for many years.

The three drawings below were done by designers working with Chuck Jordan on his twelve- and sixteen-cylinder "King Car" project. Jordan's concept of the project resulted in many lasting and influential ideas, some of which carried over into later production models. Some design themes found their way to the 1967 front-wheel drive Eldorado and some later to the 1971 Eldorado. Wayne Kady drawings are noted, though he may have penned some of the others as well.

The name "Cyclone 16" appears on the front fender of this Wayne Kady drawing.

Kady did this drawing in December 1963. The blade fenders and the rear-end treatment are similar to the 1967 production Eldorado.

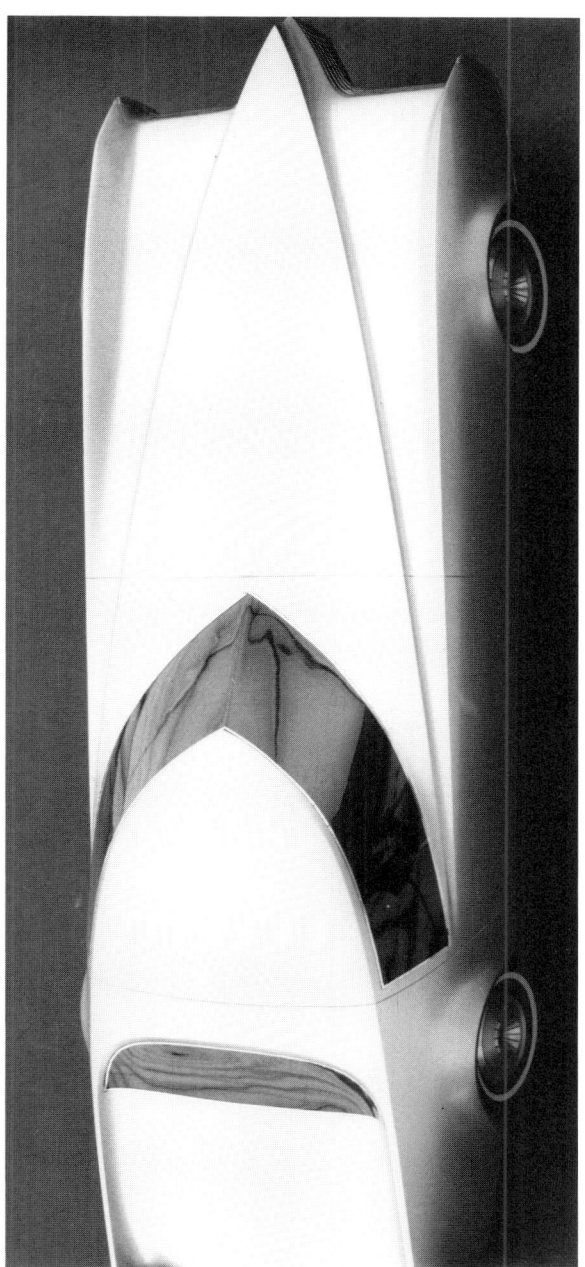

Above and next page top, a scale model from the twelve- and sixteen-cylinder project. Obviously less expensive to build, scale models of some of the designs were built and photographed. For some reason we have only been able to find examples of three sportier designs. At first glance this design looks as though it might work for a Corvette, but the hood length is extreme. Blade fenders and a pointed windswept look make up the overall uncluttered design of the car. It appears that entrance is by lifting the windshield/roof assembly and stepping in. Again, the rear corners of the car are similar to the production 1967 Eldorado. Some of the photographs were made in front of the Design Staff Building's fountain at the Tech Center.

A more conservative scale model from the twelve- and sixteen-cylinder project, this design incorporates many of the features of the 1963 Cadillac in the frontend and the rear fenders. The fastback and elongated hood shows the design's kinship to the other models.

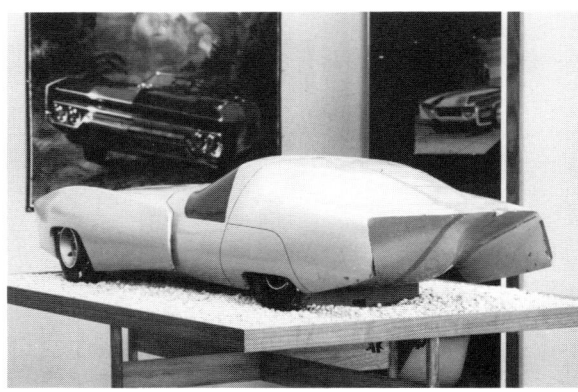

Right, this scale model became the XP-840 and was later built in a full-size fiberglass model. Although a radical design, the blade front fenders and the trailing edge of the rear fenders provide a couple of cues for things to come. The almost separate front blade fender appears in a much more conservative and dressed form in the 1971 Eldorado.

Above and right, one step in the design process, some variations on one theme. In a more conservative way, a simple personal car design was tried. The simplest form of the theme is on the left: one crease on the side and an uncluttered front-end. A four-door version is shown on the right, from a front three-quarter view. Here the car has a coffin-nose theme.

Above and left, evidently, Pininfarina was asked to provide design suggestions for the concept shown in the previous group of photos. Here are photos of two models bearing the Pininfarina insignia just behind the front wheel and presented in July and August 1963. Pininfarina's first model is shown on the left in rear three-quarter view. Here the crease is emphasized by a band of chrome that almost encircles the car (it does not cross the rear end). "Cadillac" is spelled out on the left side of the rear deck. Notice the 1963 or 1964 Buick Riviera model in the background. The second model, produced a month later, is simpler and less cluttered with a crease running the length of the rocker panel added.

1966 Cadillac Fleetwood Eldorado convertible production car. GM's October 6, 1965, afternoon press release for the car read, "The 1966 Cadillac styling story is highlighted from the rear by a redesigned deck lid, new lighting in restyled bumper outers, and a new rear bumper with the lower half painted in body color. A striking new front appearance is achieved through the reduction of chrome and particular attention to lighting detail. It remains unmistakably Cadillac, retaining the traditional cross-hatch grille design." The horizontal center bar across the front grille, which included the parking lamps, distinguishes this car from the 1965 at a quick glance.

continued from page 75
First Front-Wheel-Drive Fleetwood Eldorado

Some people mistakenly call the 1967 model the first Eldorado, but this car would not have been developed had it not been for the marque's history of innovation and design excellence. The car represented a return to the idea that an Eldorado should be unique and truly special. Since Italian production of the Eldorado Brougham had ceased during the 1960 model year, Eldorado had been only a specially trimmed Cadillac. The time had come to set things straight. Luckily for admirers of fine automobiles around the world, Cadillac Division decided to return Eldorado to its former position of having a distinctly separate body, and they have not altered this decision since.

Car and Driver actually called the 1967 Fleetwood Eldorado "a high-roller's version of the Oldsmobile Toronado!" That could not have been further from the truth. The Eldorado was not a badge-engineered reclothed Toronado. The front-wheel-drive Toronado had been introduced the year before, in 1966, amid much fanfare. GM could have presented the Eldorado simultaneously but did not because the people in the Cadillac division, especially the engineers, did not feel their car was ready. The Eldorado had profited from the combined engineering efforts of the specialists at Cadillac, Oldsmobile, and Buick to make this car possible, but the Fleetwood Eldorado was certainly a cut above the rest.

GM had tinkered with front-wheel drive a bit in the thirties. The system had even been considered in the LaSalle II show car at GM's 1955 Motorama. (When the LaSalle was built, it was powered by a V-6 and rear-wheel drive through a transaxle.) But it was not until Dan Adams (assistant chief engineer at Cadillac) and Tom LaRue (assistant sales manager at Cadillac) began to work out some ideas for a replacement for the 1959–1960 Eldorado Brougham that things actually got moving. By 1959, Cadillac had tested a front-wheel-drive vehicle at the GM test track. Oldsmobile did not test such a system until the following year.

Dan Adams had surveyed the world automotive manufacturers concerning front-wheel drive. No one had given him much encouragement. Indeed, the Europeans were of the opinion that front-wheel drive just would not work on any car with an engine of more than two liters.

The Eldorado's stunning design had developed through a series of styling projects. Chuck Jordan's idea of reaching for new ways to express the elegance and sportiness of what he thought Cadillac and Eldorado should be, culminated in a massive creative output evolving through three or four avenues of development. Several top designers, including Stanley Parker, Wayne Kady, Jerry Brockstein, and Chuck Jordan, were directly involved in contributing to the final outcome over a period of years. This, of course, developed under the watchful eye of Bill Mitchell, who was not one to hide his feelings about what he wanted a car to be.

The 1967 Cadillac Eldorado would be a unique automobile, but one must remember that there were parallel development programs going on at the same time at Oldsmobile and Buick.

As the new Chief Designer in Cadillac Studio, Stanley Parker finished the exterior design, producing a fiberglass model by May 1964. George Moon had been in charge of the interior design and had created the instrument panel and dash. Approval and development filled up the time from this point until the introduction of the 1967 production model. There were a great many hurdles to overcome.

Assistant Chief Engineer Dan Adams was told by many people, for example, that the Morse Hy-Vo Chain (a division of Borg-Warner) mechanism that was designed to transmit the power from the engine's flywheel to the Turbo Hydra-Matic transmission (which was mounted parallel to the engine) just would not hold up. These mechanisms have proven to be quite reliable.

Some automotive forecasters had said that Cadillac

1966 Cadillac Fleetwood Eldorado convertible production car. This is one of the most simple and elegant designs to come out of Cadillac Studio. All Eldorados for the year came with triple-stripe whitewall tires and bucket seats were again a no-cost option. This was the last of the rear-wheel drive Eldorados and the last of the Eldorado convertibles until a front-wheel drive version was introduced in 1971. The care taken in designing this car can be appreciated even when the top is up. Notice how the form of the top complements the body design rather than distracting from it. Owner: John Willows.

could not introduce a separate model because its production facilities were already running at capacity. Cadillac took some of the extra time they had to lay out a completely new assembly plant on Clark Street in Detroit to produce the new Fleetwood Eldorado.

Finally, the Fleetwood Eldorado, with a one-inch longer wheelbase than Toronado, was introduced for model year 1967. However, the fanfare for its presentation to the public was much less than that of the Toronado's debut the year before. In 1966, the public had rushed into Oldsmobile showrooms to marvel at the Toronado's flat, humpless, floor–a welcome side effect of front-wheel drive. In Memphis, Tennessee, Elvis Presley had run into a showroom to buy a Toronado off the floor, only to find it had already been sold to a local cardiologist.

Compared to these opening day events, the Fleetwood Eldorado's introduction was low key. But its effect on the public was lasting. Cadillac wanted an atmosphere of scarcity to accompany the car's first months on the market.

Famous Milwaukee-based industrial designer and automotive enthusiast Brooks Stevens read about the car and sent all the way to St. Louis so he could buy the third one produced, for his wife. The car was met with a great deal of respect, and it turned heads wherever it was driven.

John R. Bond, publisher of *Road & Track* at the time, said, "Not since the V-16 has Cadillac produced an automobile so intrinsically attractive to the car enthusiast." This was an automobile that caught everyone's imagination. *Automobile Quarterly* awarded the car the 1967 Design and Engineering Excellence Award.

This was not just another pretty face. Optional vented-caliper front disc brakes, a torsion-bar front suspension, and automatic level control chassis made this beauty a pleasure to drive. Brooks Stevens' wife, Alice, deemed this her favorite car to drive. That was quite an endorsement from a woman who had a choice of 120 cars available to her, everything from Excaliburs to Lincolns.

The most magnificent accomplishment the designers had achieved in restoring the integrity of the Eldorado name was that they had produced a unique automobile that was—despite its extraordinary form—still a Cadillac in every way. "Instantly recognizable as a Cadillac," wrote *Motor Trend* at the time.

Cadillac had a huge investment in this car. The die used to stamp the huge rear quarter section was in itself horribly expensive to produce. Other manufacturers would have cut corners. Cadillac did not. The result was a winning design that started making money and admirers the moment it went on the market. In 1966, Cadillac had sold 2,250 Eldorados at $6,631 apiece. In 1967, Cadillac sold 17,930 models at $6,277 each. From there, sales rose to the 23,000-plus level through the various facelifts until the model run was ended in 1970.

The Toronado had a good first year, but really did not do any better than Eldorado after the front-wheel-drive Cadillac was introduced. Even though the Cadillac was priced more than $1,000 beyond the Toronado, once the buyer had seen the quality and design built into the Eldorado, he or she might have had second thoughts about buying the Olds.

Minor Bodystyle Changes 1968–1970

For 1968, the Fleetwood Eldorado had its hood stretched 4.5 inches to make way for GM's new hidden windshield wipers. The parking lamps were moved to the leading edges of the front fenders. Small side-marker lamps with tasteful wreaths and crests were mounted on the rear fenders. A 472ci V-8 was the power under the hood.

In 1969, the hidden headlights were abandoned in favor of exposed headlights mounted in the car's front grille.

The Eldorado name in block letters returned to the front fenders, just behind the wheels, for the 1970 model year. On the left side of the front grille, an "8.2 Litre" badge announced that a new 500ci, 400hp V-8 engine was now the powering the model. Thinner vertical rear lamps, vees on the front parking lamps, and chrome rocker panel trim were all part of the car's new appearance. For the first time, an electric-powered sunroof was optionally available for an Eldorado.

The Eldorado concept had been reborn in a big way. The enormous success of this great American car would make it difficult for the division to return it to the status of a specially decorated Cadillac.

Chuck Jordan Interview

1959–67 Eldorados, 1971 and 1979 Eldorados, 1986 Eldorado, Twelve- and Sixteen-Cylinder Projects, and the Current Eldorado and Seville

Chuck Jordan joined GM Styling in 1949 after graduating from MIT in automotive engineering. Began his career in truck, tractor, and train design, until one day when Bill Mitchell called him into his office and said, "Kid, if you're going to get anywhere around here, you've got to get into cars." Jordan was assigned as chief designer of an Advanced Design Studio, his first car being the Buick Centurian Motorama show car. This was followed by the 1958 Corvette. He was promoted to chief designer of the Cadillac Studio in 1958 and joined the studio as the 1959 Cadillac program was in progress. Jordan headed the Cadillac Studio until 1964 when he became Bill Mitchell's assistant, watching over all GM exterior design. He went to Germany as director of Design for Opel between 1967 and 1970. After his return, Jordan took on the responsibilities of executive in charge of Exterior Design. In 1977, he was named director of Design. In 1986, he was elected vice president of General Motors Design. Jordan retired in 1992 after forty-three years in Design.

Author: So you've always had a love for Cadillacs?
CJ: Yes, ever since my dad bought a '48 fastback coupe. I loved that car. It was the first Cadillac with fins. The lines were beautiful!

Author: But the first full Cadillac design program you worked on was the 1960?
CJ: Actually, when I joined the Cadillac Studio they were in the middle of the '59 program. So we completed the '59 and started the facelift for 1960. The 1961 Cadillac was my first all-new program. During those days, we did some wild Cadillacs. We always started a program that way. I remember, in particular, the beginning of the '61 program. In fact, I still have the original sketch that started it all—a Coupe DeVille with an upper fin and a lower fin, both right over the rear wheel. We made a full-size clay model of that design. It was distinctive, to say the least! Fortunately, sanity prevailed and we focused the ideas into a more appropriate design where the lower fin ran the full length of the body as side protection. In 1963 and 1964, the fins became much less prominent, and we returned Cadillac design to a simpler, more elegant and distinctive form.

Author: During the early sixties, the Cadillac Eldorado was not a separate model from the rest of the line. What was your philosophy behind the Eldorado? Did you think of it as a trim package or as the flagship for Cadillac?
CJ: We always thought of the Eldorado as the sporty flagship for Cadillac. But, in actual fact, all we were allowed to do was trim. Looking back, I think what we did with the Eldorado trim made a lot of difference—the Eldorado always looked more elegant and more expensive than the rest of the line. But to us, trim was not enough. We needed to design a real flagship: a Cadillac that was distinctive, sporty, and personal, and stood on its own.

Chuck Jordan.

Author: Is that how the '67 Eldorado came about?
CJ: That's right. We believed strongly in our philosophy and we did something about it. Toward the end of 1959, in the back room of the Cadillac Studio and sometimes in an advanced studio, we experimented with "sporty elegant" design directions. This went on until about the end of 1964 and resulted in a number of scale models and several full-size clay models which we reviewed with Cadillac at regular intervals. These models created a great deal of interest and excitement, but the timing and financial situation never seemed to be right for a production program. You know, timing is everything! It wasn't until the Toronado/Riviera front-wheel-drive program materialized that we saw an opportunity to do this Cadillac we wanted—that turned out to be the '67 Eldorado.

Author: Who do you consider to be the primary designer of the '67 Eldorado?
CJ: You know, in our business no one person ever designs the whole car. It's a team effort. In the case of the '67 Eldorado, there were a number of designers involved during the years of advanced design and some who contributed to the production design. They all deserve credit. Some names come to mind: Ron Hill, Stan Parker, Jerry Brockstein, Don Roper, and Wayne Kady. I was also very much involved. But let's not flatter ourselves. There was only one guy who was boss, who set direction and made decisions—and that was Bill Mitchell. If he liked what you were doing, fine. If he didn't like it, pow! Bill Mitchell, as vice president, was responsible for designs and, believe me, he took that responsibility seriously!

Author: I've heard about some V-16 designs that you guys were doing. Tell me about that.

CJ: This was kind of an extracurricular "fun" project that had to do with the idea of a two-passenger luxury Cadillac built around a V-16. It was strictly a designer's dream—an image study. We were fascinated with the impact a sixteen-cylinder Cadillac would have on the Cadillac image and we actually completed several scale models to make the point. These designs were spectacular! Now that I look back, they were almost cartoons. But that was the era of long hoods, and boy, did these cars have long hoods! At one time we considered building a running car. That proposal used some of the '64 sheet metal, moved the driver rearward near the rear wheels, and incorporated two V-8s attached together to create a V-16. Can you imagine that car driving down the street? We never actually got around to building a running car—and maybe that was a good thing—but we sure made our point. Later, we designed another two-place car with an advanced body shape. That design was significant and had an influence on what we finally did leading up to the '67 Eldorado.

These fantasy projects are good "creative exercise." They're exciting projects you think about while you're driving home from work. They educate management about the possibilities, and they have real value for a designer when he finally has to design a car under production conditions. Besides, they're fun! If you're not having fun in the design business, you're not doing it right.

Author: How did all this advanced work and these "fun" projects finally evolve into a production Eldorado design?
CJ: When the specifications for the front-wheel-drive platform [Toronado and Eldorado] were established, the Cadillac production design was assigned to the Cadillac Studio. If you study photographs of our advanced work, it's pretty clear that the theme for the production design was well defined. We didn't have to scratch our heads and ask ourselves, "Now what do we do?" We'd had enough practice to know exactly where we were going and we executed the production design without a hiccup. I think the '67 Eldorado was one of the most beautiful cars of the era. As designers, we were happy to finally get a sporty, elegant Cadillac on the road.

Author: You were in Germany at Opel from '67 through '70. What happened when you got back?
CJ: I missed out on the '71 because I was gone. That was the last of the "big ones." I happen to know this was Bill Mitchell's favorite. The '79 was the first of the downsized Eldorados. I missed out on the '79 development also because I was watching over Chevrolet, Pontiac, and commercial vehicles at the time. But I remember the era well. Those were the days of "downsizing." That meant, to meet government fuel economy regulations, the weight of the car had to be reduced and, to do that, the car's size had to shrink. Nineteen seventy-nine was the Eldorado's turn to be downsized. For the designer, downsizing was a "cultural shock." Instead of following what we'd always been taught "lower, longer, wider," all of a sudden the rules changed. Now it was "shorter, narrower, and higher."

Downsizing an emotional car like the Eldorado was particularly difficult. The '79 took a lot of extra work to tailor the design to the new dimensions, but I think the guys did a good job and the car came out with Eldorado character that looked tighter and more efficient—not so excessive in its proportions as in the past.

Author: What about the '86? That design seemed to have lost something.
CJ: You're right. It was certainly not an impressive car. Some said it wasn't even an Eldorado. Irv Rybicki was vice president (having taken Mitchell's place when he re-

A sidetrack in the design story: The XP-840 full-size fiberglass model of a late Eldorado concept was presented in December, 1965. That this design was a serious consideration is seen in the "ELDORADO" logo above the front bumper on the driver's side. Note the periscope arrangement on the roof necessitated by the lack of a rear window.

tired) and I was director of Design. We were living through the second generation of downsizing. Those were tough days for the designer. I remember the Cadillac chief engineer insisting, "If we're going to err, it's going to be on the side of smallness"—and we did! I don't have any excuses. Under the restrictions, we just couldn't get the right proportions or much emotion in the design. We did the best we could, but it sure didn't turn out to be a classic.

Author: What about the Eldorado on the street now? How were you involved in that?
CJ: In October 1986, I was appointed vice president of Design. The first day I got that job, I called the whole staff together (about 1,300 people) and talked about some goals for GM Design that I believed in strongly. I talked about bringing back the excitement and flair in our designs, about strengthening the divisional image —no more look-alike cars—about designs that were distinctive and had that intangible quality called "beauty," about designs that would move people emotionally. And I added, "Let's have some fun doing it!" and we did, as it turned out.

When the time came to design the next generation Eldorado/Seville, we were ready. Things had changed since 1967. The market now was much more complex and sophisticated—it was international in taste and character. In meetings with Cadillac Marketing, Engineering, and Manufacturing, and most of all, Cadillac Management, it was clear our new Eldorado and Seville had to be exciting personal luxury cars that met international standards.

The design of the Eldorado was started first—months earlier than normal—so we'd have time to experiment. This time around, the Eldorado had to be right, no excuses. We decided to get three teams going at once. We needed some different options to look at.

In the Cadillac Studio we concentrated on a design that projected the Eldorado heritage: what you'd expect the new Eldorado to be. In an advanced studio, we took more risk and focused on a racier, sportier design. Then, to get an international perspective, we asked Sergio Pininfarina to design an Eldorado the way he saw it.

Those were exciting days! All the designs were over the same architecture, but each developed its own individual character and appeal. Each design had its strong points and its weak points. But when we brought them all together for review, not one of them screamed, "I'm it!" So, for the final design, we distilled the best of each into one design that was better, as it turned out, than any one of the original proposals.

Author: What happened to the Seville? Wasn't it done at the same time as the Eldorado?
CJ: The Seville development is an interesting story. We had been trying so hard, grunting and groaning, on the Eldorado design, it suddenly dawned on us that we hadn't done anything about the Seville. I got hold of Dick Ruzzin [chief designer of the Cadillac Studio] and asked him to get going posthaste.

He did! Several weeks later, after coming back from a trip, I went in on a Sunday to walk through the studios. When I opened the door to the Cadillac Studio, I stopped in my tracks. There in front of me was a new Seville design done in clay-colored cardboard and black tape pinned onto a full-size clay model. I looked at it and said, "That's it!" The final Seville design really never deviated from this "cardboard and tape" proposal.

There was a lesson to be learned here. The Seville, in haste, was done with spontaneity and flair (like a sketch). On the other hand, we had been working so hard and long on the Eldorado, the design somehow looked frozen and worked over; so just before the final surfaces were finished, we gave the Eldorado a "cardboard and tape job."

That simple exercise put some spontaneity and flair back into the Eldorado design. Finally, we all felt, "Now we really had an Eldorado and a Seville we could be proud of."

Lesson One: "You can't schedule creativity." Lesson Two: "Don't quit 'til it's right."

Author: Can you give us an idea of what the next Eldorado will be?
CJ: After the new Eldorado/Seville came out, I chose an Eldorado Touring Coupe as my personal car. It was clear from the beginning that the STS was stealing the show with its dash and understated no-chrome design. So I sent my Eldorado to Heinz Prechter at ASC [American Sunroof Corporation] and asked him to "STS" it and paint it a dark wine color. He did and that car caused a sensation wherever I went. That's a clue to the next Eldorado.

Author: What do you think is the future of the Eldorado?
CJ: I'm retired now, so I really don't know what's going to happen to the Eldorado past 1996. But that doesn't keep me from having an opinion.

Consider this: In today's market, the popularity of four-door cars is increasing. There are some great handling and great performing luxury four-door cars out there right now. A few of them, like the STS, have a sporty emotional magnetism that rivals any luxury coupes on the market. So it's no surprise that the popularity of luxury coupes is tending to diminish. This makes life tougher for the Eldorado.

These are the facts. But deep down and on an emotional level, I believe the Eldorado will continue to be a flagship car for Cadillac if it's a strong leader in the international market and if its design "makes your mouth water."

I have faith in Cadillac. In all my years working with Cadillac, I have never worked with a more competent, receptive, forward-looking group who knew where they were going and how to get there than the Cadillac team on the Eldorado/Seville project. John Grettenberger, general manager of Cadillac and one of my favorite people, was the keeper of the vision and, like the "Pied Piper," he led all the disciples toward a common goal. It was a great experience to work with John and the Cadillac group. Some of the players have changed now, but I have complete faith in John Grettenberger to lead Cadillac to an even stronger position. He and his team will do the right thing, whatever it takes—and I think that includes a new Eldorado.

Stan Parker Interview
1960 and 1967 Eldorados

Parker started at GM with Buick Studio under Ned Nickles in 1953. After military service, he returned to Buick Studio as assistant chief designer before going to Cadillac in 1962. Working as Chuck Jordan's assistant, Parker helped Jordan in his attempt to bring elegance back to the Cadillac line after the 1959 model. In August of 1962, Parker was made chief designer of Cadillac Studio, a position he held through 1968. He worked on the first front-wheel drive Toronado, introduced in 1966, and worked as primary designer of the 1967 front-wheel drive Eldorado. Parker served as chief designer at Advanced Cadillac Design Studio from 1972 through 1980, and he was responsible for the 1975 Seville. He was chief designer of Studio 2000, then chief designer at Advanced Vehicle Concepts Studio ("Dream Work Shop") from 1980 until his early medical disability retirement from GM in 1987. Parker designed the "Lean Machine" concept vehicle shown at EPCOT Center and used in the movie Demolition Man.

Stan Parker.

Author: Since you worked on the '60 Cadillac models, do you remember any specific parts of the design that you worked on?

SP: That tailfin took a couple of days to develop. We wanted to reduce it from what it had been in '59—you remember the '59's big fin with the bullet taillights mounted on it? I would do a sketch of the rear fender (a full-size side view on vellum paper) and Chuck Jordan would have to leave and then he would come back in about five or five-thirty and say, "No, I'd like this and this changed—just a little bit here and there." I don't know how many rolls of vellum I went through, but after two days he approved the profile. That was what ended up on the car.

That experience entertained the whole studio staff. There were lots of laughs, and I joined in them.

Author: You received the photograph of master industrial designer Brooks Stevens' treatment of his '67 Eldorado?

SP: Yes. It was interesting that it is a little similar to the show car for a national auto show we did for the introduction of that model. We had a different kind of rocker treatment, an open top over the front seats, and a line, similar to the one on Stevens' car, reaching from the crease of the hood back along the beltline.

Author: Do you have one of the '67 Eldorados now?

SP: No, but Wayne Kady, who worked with me on the design, has one. He brought it by my home recently. He did all the painting of it himself. I hadn't realized he had that skill. It is just beautiful.

I did have the number seven production car for '67, but I didn't keep it because I thought that I'd eventually design a car ten times better.

Author: The '67 design started in an advanced studio, as I understand it.

SP: Yes. It started in an advanced studio and from there was moved to the Cadillac Studio. Wayne Kady worked with me on it as my assistant. I had been working on the '66 Toronado before that, but we had only gotten to the point of laying out the driveline and all that. My only real contribution to the '66 Toronado was the front end. When I was leaving the studio, Advanced Number Three, I left the sketch for the front end on a wall. That was what ended up on the car. I wish I still had the sketch in my files.

Author: What was the input from Bill Mitchell and Chuck Jordan on the '67 Eldorado?

SP: There was very little interference from Mitchell or Jordan on that car, which was a surprise for me. Every square inch of that car, I had my finger on. The only major contribution that Mitchell made had to do with the car's fins. Originally there was a fin from the rear quarter blended in with the sail panel, like the one I later put on the Oldsmobile Cutlass Supreme coupe. Mitchell said to hollow out the fin between the sail panel and the outer surface so that it stuck up like a fin rather than being a paste-on. We followed his suggestion.

Fisher Body had a hell of a time making that quarter panel. Finally, they had to make it in one piece. It went from the door cut to the little quarter panel window, up the sail panel, and the joint went up a little before going down to the backlight. They put the actual deck cut on the inside of the quarter panel, which went all the way under the car, then went back up to the rear wheel opening, and back to the rocker panels at the bottom of the door cut. That was all one piece.

A body engineer from Fisher Body called me one day and asked me to come over to see what they had to go through to build the quarter panel. He almost called me a

The mainstream design path: developing the XP-825, the immediate precursor to the production 1967 Eldorado.

name! Anyway, when I got there, he took me into an experimental room where they had this huge steel ball which was part of the die. This ball drew the metal in behind the profile of the fin. He said, "We are having to throw away twice as much sheet metal as goes into the quarter panel, every time we make one." It was the most incredible thing I ever saw. I said, "All you have to do is plug in the light and you've got the whole rear quarter here from one stamping!"

I said to myself that this is going to be quite a job for a repairman to work on one of these quarter panels. Being a one-piece stamping, where do you begin?

Author: Do you have any idea why the '67 style model only lasted through 1970, whereas the '71 style model run went on for several years more than that?

SP: Yes, part of the reason had to do with the difficulty of making that rear quarter panel. If you wanted to make substantive facelift changes, say by changing that quarter panel, you just couldn't do it. For example, Cal Werner would say, "Let's make the car look different by taking the end cap off the fender and putting in a turn signal light." I remember that when Werner took over as Cadillac president and general manager, he said he didn't think the turn signals should be in the bumper anyway. After that, he had the concealed headlights removed to save money. With that car, anything you did much beyond that would deteriorate the design. It was expensive to produce the tools and dies to build that car. GM and Fisher Body had to do all kinds of special things to get it into production—even the hood had to be produced in two pieces.

But this was a different time from the way things are now. Nobody ever said, "Hey, we've got a winner like it is for '67, let's just keep making it the way it is." Today, with the Saturn, if they sell it for five years, nobody cares if there's no significant styling change. That's the way the Japanese operate too, as long as people buy it, let's not change it.

Author: That would explain it. That rear quarter die, in particular, was so expensive it was difficult to make a change.
SP: Yes. After that model had been in production awhile, one day Mitchell called and asked me to stay around a little after work. He wanted to see me. It was on a Friday. On that particular day, we had just found out that Cadillac had broken all previous sales records, having sold over 200,000 cars for the first time—213,699 to be exact. So, I thought to myself that Bill Mitchell was coming down to tell me he was going to give me a big pay raise.

It got to be 5:30, most people were gone and half the lights were out. Finally he showed up. He said, "Come on, I want you to meet somebody."

We went down the hall to Oldsmobile Studio and went inside. All the lights were out except for one spotlight on a coffee table in the middle of the room. Mitchell said, "Stan, I'd like for you to meet your new bosses." The two men he introduced me to were the heads of Oldsmobile.

I said, "Do you mean I'm going to be designing Olds-mobiles?"

"That's right; I've gotta go," Mitchell said, and he left me alone with the two men.

So these two guys, Metzel and Beltz, then tell me that since I've done such a good job in Cadillac, they hoped that I can do the same thing in Oldsmobile. They were in ninth place at the time. "There's just one thing," they said, "we can't afford bumper blank sizes the way you had them at Cadillac. You've got so much metal in there, it's too costly. Then you've got all those expensive die-cast taillights and other expensive stuff like that; we can't afford to do a car like that. But we'd certainly like you to try to get us out of the sales position that we're in—without spending as much money as you did at Cadillac.

I said, "Well, I believe in miracles, but there's only one other guy I've heard of who could walk on water and I don't know if I can do that. If you'll excuse me, I've got a weekend trip planned with my family so I'll probably see you on Monday."

Author: Could you tell me something about the wheel treatments on the '67? I know that what was produced was unusual for the time.
SP: Suddenly, wheels were a big thing. I remember we were doing some cast aluminum wheel designs. I remember the guys were doing a ton of them, getting them fiberglassed and then chrome-plated. This was going far beyond just having a hubcap. We were getting into the functional part of it. We even tried to get cooling areas in the wheels.

That reminds me of an interesting story concerning the '67. When we first had the fiberglass model of the Eldorado, we presented it to Cadillac out on the patio. The car was painted with actual gold flake paint. I don't know if it was 24 carat or not, but I wouldn't even venture a guess as to how much it cost. When that gold car was taken out into the sunlight, it really looked like something.

We got some of the guys in Product and Exhibit to help us in designing a presentation kind of thing, rather than just putting it on a turntable on the patio. You've probably seen pictures of cars on the patio. We got tons of golden asters, the big flowers they have in football games, all over the place. We had an arched trellis—big enough to walk through—in the background. You'd walk through the arched trellis, with all those flowers all around, and there was the car. The guys did a fabulous job.

Harold Warner of Cadillac Division said, "Oh, my God, we've gotta get the money to build this car!" Everybody agreed.

Warner told everybody that was with him from the division to get back to Cadillac and start going through the books because, he said, "We've gotta build this car!"

Eventually, we had to show it to Jack Gordon and Frederick Donner, the CEO and president. I had always called these guys "Donner and Blitzen." Anyway, we did the same thing, only a modified version of the display we had used outside, on the turntable on stage in the GM Tech Center auditorium. There were just a few people there, Bill Mitchell and some other executives. As vice president for Design, Bill Mitchell presented the car to the other executives.

Donner said, "Bill, there you go. You're putting that vertical line on that quarter window. You know how I feel about that. I hate vertical lines on the quarter window! Would you think about putting a slant on it?"

Mitchell said, "No, this gives you more privacy in the back. It gives the impression of exclusiveness for people riding in the vehicle. Besides, it goes with the whole design of the car."

Donner said, "I never did like those vertical lines. I'll think about it."

When they walked out, I told Wayne Kady, "Below this stage is a storage area. I want you to get the union movers and clear out the very back. The area makes an L-shaped turn at the back. Pull all the garbage that's there out, and move this car in there. Put it way in the back and put an old dirty drop cloth on it. Then put all the garbage back in there in front of it and conceal that car. That's where we'll keep it until somebody maybe wants to see it. We'll just wait them out."

Later, but all of a sudden, the car went into production with the vertical quarter panel window—it was on the road. On August 27, 1967, I got a telegram from Automobile Quarterly saying that the car had been awarded the 1967 Engineering and Design Excellence Award. They said that only two other automobiles had been given that award in the past five years.

The day I got that telegram, Bill Mitchell came into the studio with Chuck Jordan. By that time we had the fiberglass model back in the studio. Mitchell said, "Chuck, if I knew that Cadillac Division would come up with the money to build that car, it certainly wouldn't look the way it does now. I'd have changed it!"

I said, "Bill, I just got a telegram from Automobile Quarterly, giving the car the 1967 Engineering and Design

Excellence Award."

He turned a vivid red. I thought he was going to blow up. He and Chuck left the studio.

Author: One thing I thought was so great about that car was that it was so different from the Toronado; the design really said, "Cadillac." The Toronado was a nice design, but it was a heavy-looking car, while the Eldorado was light and elegant.
SP: Right. I didn't like the fastback and the sagging lines in the profile of the '66 Toronado.

Author: Mitchell told me one time that he had really liked the Toronado until one day when he was in New York City and saw one go around a corner. "Gosh," he said, "that thing is just too big!"
SP: Well, the problem is that you might say the '67 Eldorado is too big now. I remember when Wayne Kady brought his over to my house. I told him, "This car didn't look this big when we did it in the studio."

Author: But the Eldorado's design is very fresh. If someone saw one parked on the street today, you might think it was a new car. It hasn't been copied that much.
SP: Later, after Oldsmobile, Wayne Kady and I traded places and I went into Cadillac's Advanced Studio, while Wayne became head of Cadillac Studio. That's when I did the '76 Seville which got the Car of the Year Award from Motor Trend.

Bill Mitchell's personal version of the 1967 Cadillac Eldorado. Quickly identified by the huge lights built into the bottom portion of the bumper, this was the 1967 Cadillac Eldorado Mitchell had specially built for his own use. Mitchell's specially prepared cars were not frivolous toys, he used them as mobile laboratories for both design and engineering development.

1967 Cadillac Fleetwood Eldorado, production model. The long and involved creation process was worth the effort. Introduced a year after Oldsmobile's front-wheel drive Toronado, Cadillac's engineers knew their product was right and ready for the road. Built on its own production line with an exclusive 120in wheelbase, the Fleetwood Eldorado was shorter than the standard Cadillac and was the least expensive Fleetwood Cadillac offered. This was the first front-wheel drive Cadillac and demand for the 4,500lb beauty resulted in sales of 17,930 copies. Although modified for front-wheel drive, the car was powered by the same 340hp, 429ci V-8 that powered the other 1967 Cadillacs. The long hood and short rear deck were an instant hit with the public, and the unique design made the car stand out wherever it was seen. Note the creased rear window. The slotted wheels were something quite new at the time, giving the car presence and calling attention to its movement. The sharp corners and creased sides made this a difficult and expensive body to produce, but Fisher Body and Cadillac went the extra mile to bring the car into production with practically no compromise to the realities of bending metal. All this was done while retaining enough design cues that everyone recognized the car instantly as being a Cadillac.

The 1967 Eldorado's primary designer Stan Parker is shown here posing with GM Designer Wayne Kady's car. Kady worked with Chuck Jordan on the twelve- and sixteen-cylinder project and with Cadillac Studio Chief Designer Stan Parker on the design of the 1967 Eldorado. Kady credits Parker with being the primary force behind the design. The car shown here was Kady's sister's car at one time. He acquired it from her and restored it to its present pristine condition.

George Moon Interview
GM Tech Center History, the 1967 Eldorado, and Later Eldorados

George Moon was responsible for the 1963 Buick Riviera interior as chief of Buick Interior Design Studio. He was primary interior designer for the '67 Eldorado as chief of Cadillac Interior Design Studio, and as head of all interior design at GM Design Staff, he oversaw all Eldorado interior designs until his retirement. Moon currently manages his own design firm and is writing a history of the General Motors Technical Center.

Author: I understand you are writing a book on the history of the GM Tech Center. Do you know which of the Eldorados was first designed there, rather than in the old building in downtown Detroit?

GM: The keys were handed over to the styling people by the building contractor on September 15, 1955. Work began right away. Of course, there was some carryover from what had been done in the old place, but I'd say that the first Cadillac that was completely done at the new Tech Center was the '59. The Cadillac design for that model year, and the other divisions too, was scrapped when we found out what Chrysler was doing with those low roof lines and big fins. I never will forget that day they drove a Chrysler, a DeSoto, and a Plymouth into the garage at Design. Harley Earl went berserk! He couldn't get over those fins! After that, the design that became the '59 Cadillac was started.

Author: Could you tell me about your experience with the Cadillac Eldorado?

GM: Yes, I was chief designer in the Cadillac Interior Studio up until mid-1968 when I was sent overseas with Chuck Jordan to Opel. But from '63 to '68 I was in Cadillac and we began the interior design work on that all-new Eldorado (the front-wheel-drive '67) in about '65. I had also done the interior for the Riviera [1963] in Buick. Of course, that started out as a Cadillac, but ended up a Buick.

George Moon.

Author: What year did you retire from GM?

GM: 1987. I was involved indirectly with Cadillacs after the work on the '67 because I had all the interior studios as head of all interior design for GM. But when you're managing the whole group, you're a few steps removed because you're overseeing everything that goes on in the interior studios. You make some decisions, but you leave most decisions to each chief studio designer to execute the IPs [instrument panels] and the total trim. But we did all those Eldorados. The last Eldorado that I was involved with was probably in the early '80s when we were starting IPs and trim and so forth for the interior— what would come out in '84 or '85. Those would have been the last cars that I touched. I did do some Eldorados in the Advanced Design studios. When I went into Advanced, we did concepts for Eldorados. One of them became a Cadillac show car, actually we did a couple of show cars. The designs never manifested themselves in production, of course.

Author: Do you remember their names?

GM: Well, we had the Voyage and then we had the Solitaire. One of the things we had in that was the glass below the beltline. Actually, the car was designed by Alan Young in our Advanced Studio Number Five. He had done this beautiful monocoque shape. The glass was split at the beltline with a thin body color panel running through it. We had glass below the belt as well as above it.

Author: I saw the Solitaire at Motorama in Houston. It was beautiful.

GM: The Solitaire was an Eldorado—an Eldorado theme idea which never really went further. After that, we did the Voyage (designed by Jerry Brockstein), which was more of a Seville. Those were the last cars I worked on directly.

Vince Kaptur Interview
The 1957, 1967, and 1979 Eldorados and More

After graduating from Michigan College of Mining and Technology, Vince Kaptur served as a U.S. Navy fighter pilot in Korea before joining GM in 1953 as a Styling Engineer. He spent his entire career at Design Staff, working in Engineering and Design Development, and for a period as an assistant to Bill Mitchell as his "Chief Engineer." Kaptur retired from GM as Director of Automotive Engineering and Development in 1981.

Author: As I understand it, when you joined GM you were following in your father's footsteps.
VK: That's right. My father had been one of the first people Harley Earl had hired when he came to GM. In fact, I worked with my father on some things when I first got to Styling. He was just about to retire at the time as Harley Earl's Chief Engineer.

Author: What were your duties?
VK: I started out as a Junior Layout Draftsman working for Freddy Walther. We did engineering seating bucks [seating models]. Then I worked with an advanced group [Advanced Engineering] made up of a bunch of guys who were coming up with inventions. I've got about four or five mechanical patents. I was the co-inventor of the body dimensioning system. That was my most significant invention. This was the human accommodation dimensioning system (this was a combination of a three-dimensional amorphic dummy and a two-dimensional adjustable human figure both of which could be used to obtain eyepoint positions, hip and shoulder room measurements, and other human engineering reference parameters). I also worked on the design and development of the dimensional checking devices with Mike Myal. Mike worked for me. At the time, he was a fifth year student at General Motors Institute. This system has become the industry standard.

Author: But, in general, you worked with Design Staff in turning their designs into three-dimensional, workable structures.
VK: That's right. After working with the inventor's group, I went into the Body Development Studio. That studio had many responsibilities. The first was the laying out of the basic architecture of the car. We worked with the different car divisions and Fisher Body, coming up with typical cross sections and all that business. This information was given to the design studios. Second, we'd work closely with Design Staff up to the point where we would begin our work by making master models. Our models were built because first Harley Earl and then Bill Mitchell were cautious about the accuracy of the clay models they saw in the Design Studios. So we would make master models in the Body Room.

Vince Kaptur.

Author: Were the master models fiberglass?
VK: No, they were clay. We had the largest staff of clay modelers of anyone in Design Staff.

Author: How were the master models different from the ones the designers did?
VK: They were different from the standpoint that they were accurate. The design studio models back in those days were plus or minus. The roof of a car might be cheated down. A lot of things might be different: glass development is one example. We had to work closely with our glass vendors. When the wrap-around windshield came around, it was more than drawing a picture of a windshield and modeling it up. You had to be able to bend the glass and have it optically correct. Also, the shape of the windshield determined the shape of the roof. For example, the designers might have shown a roof that appeared to be two inches thick, but by the time we got a real windshield in there, the roof might be three inches thick.

Author: Were you and your people involved in that next step of producing the fiberglass models?
VK: Yes. We had another group who did our fiberglass model engineering. We would take casts—either from studio models or body development models—from which they would lay up their fiberglass. Back then, it was all hand lay-up work. To put it briefly, the casts or molds were done in plaster and from those the fiberglass model could be made.
Another department, depending on the stage of the program, engineered the seating bucks at the beginning of the program and built the fiberglass models at the end of the program.

Author: Did you put actual glass in the fiberglass models?
VK: Yes. Our glass vendors would use this as the first chance to make prototype glass for evaluation. They

1967 Cadillac Fleetwood Eldorado, customized production model. Brooks Stevens, an automobile designer himself, instantly saw the beauty of the new Eldorado and purchased the third one Cadillac produced for his wife, Alice. He added a slash of white from the leading edge of the hood back to the door handle and the base of the sail panel. The front portion of the roof he covered with a white material that extended to the base of the "A" pillars. This was one of Alice's favorite cars, which is saying a lot, because for twenty years, Stevens and his sons produced Excaliburs and he owns a museum of more than 100 automobiles. The recently restored Eldorado was photographed in front of the Milwaukee Country Club. Owners: Brooks and Alice Stevens

would, at their expense, make up samples and furnish them for our fiberglass models.

Author: And the seating bucks?
VK: Of course, the seating bucks didn't have glass in them. They were primarily wood structures.

But we did our fiberglass models either way: with or without seats. In other words, we had Interior/Exterior Models, in which the seating was included, and we had Dummy Models, which were just purely exterior models.

Let's say we were doing a C-Body program for Cadillac. We might do three fiberglass models. We might do a full Interior/Exterior Model for the four-door sedan, so you could see the interior. But the convertible and coupe might be just Dummy Models where the doors didn't open and there was no interior. But they would have real glass in them and they looked like the real thing.

Usually the fiberglass models were built on a wooden structure. The Interior/Exterior Models a lot of times would be built on a prototype chassis, if there was such a thing available. Some of them we electrified so they would move on their own at a very slow speed. Sometimes people wanted to see the models in motion.

We had height adjusters on them so we could view the cars in loaded or unloaded situations. This was always a problem, to know how the car would look under these different circumstances. When the cars were designed in the Design Studio, they were always designed in the fully loaded condition.

Author: You had some involvement with the '57 Eldorado Brougham.
VK: I'm very familiar with Harley Earl's interactions with the design of the '57 Eldorado Brougham in the Body Room. For example, one day we were working on the roof. Earl was saying that the roof profile just wasn't what he wanted. We had worked all morning with him. Finally, he said, "We need to lower the roof one-half an inch in the back and that'll do it!"

I thought to myself, "We can't do that, we won't have any head room. We already have the man sitting on the under body right now—we can't lower the man any farther. What are we going to do?"

After Earl left, I told the guys, "On the vellum, draw the roof line a half inch higher—and then erase it out. When Earl comes in and looks at it he'll think that's where it was and that we've lowered it." He came back and he believed it. He said, "Now doesn't that look a hell of a lot better!"

I said, "Yes, sir, it does."

That's just one of the things us engineering-type guys had to go through to get a car that was practical.

At one point, in fact, in working on that roofline, he got mad. He just couldn't read the line. We'd put an overlay up, draw a new line on it, erase out the old line, and trace in the new one. When you do that, you get all these pencil marks on the backs of the drawings. In due time, usually late at night, you'd take the drawing down and erase the back to get all those old pencil marks off. But on that project, he'd stayed with us late at night and we just didn't have a chance to do it.

Well, he preceded to scold me about all those lines and how hard it was to read. I told him about not having had the time to erase the back of the vellum. He said, "With all the hell you've had to take from me over that, if I were you I'd call the company that makes the vellum and tell them to make you some opaque vellum that you can see through!"

So, over lunch, I told the guys to take down the vellum and get the back good and clean. When we got back from lunch, Earl comes back into the Body Room right behind me and walks straight over to the drawing. —And here's a pair of shoes sticking out from under the blackboard. Earl says, "What's that?"

One of the guys says, "That's our friend, Skinny, he erases the back of the drawings!"

With that, Earl turned purple. I didn't know what his reaction was going to be. I thought he was going to fire us all on the spot, but he suddenly breaks out laughing and gives me a big slap on the back. "I've gotta give you guys credit. I give you hell and brimstone all morning and you come back with a joke. That's what I like!"

1968 Cadillac Eldorado Biarritz show car. Stan Parker recalls that this car lost its rigidity when the workmen removed the front roof section. Subsequently, the rest of the body and the frame were stiffened to regain structural integrity.

But I got a hold of the guy who did it, Joe Evans, and I said, "Joe, if you ever do anything like that again, I'm gonna smash you flat, right here!" People had lost their jobs from doing less with Harley Earl.

Author: You worked on the first front-wheel drive Eldorado, the '67?
VK: Yes. A task force was set up over at Engineering Staff. Buick had determined they weren't going to do a front-wheel drive. So Oldsmobile and Cadillac had their representatives on the task force under Von Polhemus of Engineering Staff, who had spearheaded the GM development of the original front-wheel drive which involved the transfer of power from the engine to the transmission by chain. This chain linkage was a big problem, but Morse Chain (now a Division of Borg-Warner) came through with a solution to the reliability problem.

That was a very successful car. In my garage here, hanging in my museum, I have the original kick-off sketch of the Cadillac Eldorado version of the car done by Wayne Kady. It's about four feet square! It hung in my office for years while I worked at GM because I so admired that car.

Of course, in the end, this sketch looked very little like the production version of the '67 Eldorado, but it is the kickoff sketch for the project. You can see traces of the theme, but it is quite different.

This brings up something that I mentioned to you the other day when we were scheduling this interview.

Certain designers at GM said that all we did in our group was muck up their designs, but that really wasn't the case at all. We had a bunch of dedicated men there who enjoyed good looking cars as much as the next person—as much as the designers. That's why they worked there. They were the practical guys who had to make those things work. The cars had to be made buildable without ruining the general design theme.

Author: You must have worked closely with Stan Parker and Wayne Kady.

VK: Yes. I was in the Body Room at that time. We had to determine the interchangability of all the inner-body panels, the cowls, the door inners, the pillars, and the underbody. We had determined, in working with Fisher Body, the basic roof structures. Although there was maximum individual panel usage on the outer body to give the

For 1968 the Fleetwood Eldorado's parking and signal lights were moved to the leading edge of the front fenders. Red, round side-marker lights appear on the rear fender, camouflaged by the wreathed Cadillac shield. The hood was lengthened by 4-1/2in and now concealed the windshield wipers, not to mention a new 375hp 472ci V-8. Sales were a spectacular 25,528 for the model year.

individualities of the Eldorado and the Toronado, we did have to try to keep costs down by maximizing interchangability wherever else possible.

With that model, we had the added problem of performing these two functions with the E-Body which, as you know, was being used with Cadillac Eldorado, Oldsmobile Toronado, and Buick Riviera at that time. As I said, Buick had elected not to use front-wheel drive, while the other two divisions chose to do so. This meant that not only did we have to plan for interchangability between two cars using front-wheel drive, but one, namely the Riviera, that was to use rear-wheel drive. You can imagine the complications.

With front-wheel drive, there were serious problems in the development of the cowl and the dash, not to mention the front-end of the underbody. For cost reasons, we wanted to use the same cowl on both the front-wheel drive and rear-wheel drive cars. In fact, they ended up being the same, after a lot of work. The body front-end is the most expensive part of the whole car, so that's where you try to get the maximum interchangability to save tooling costs.

But, you can imagine that the humps and bumps in the floor, for example, of the front-wheel drive car are quite different from those of the rear-wheel drive car. The front-wheel drive car had a little bump in the floor primarily for the exhaust pipes to go through, whereas the Riviera had the big hump for the transmission and the driveshaft.

In addition to those problems, we got into engineering the "see-through" number one pillars that were in vogue at that time. We had been getting the preliminary rattlings of safety and vision requirements, and this was one thing that was being tried. Those number one pillars back then were made as small as we could get them—they weren't like the ham hocks in use today. I guess the ideas about that have changed. Back then, the designer didn't want to see anything there from the outside. It was all we could do to just have some molding where the glass joined the sheet metal.

Also by that time, we had gotten into the adhesive glass installation techniques, whereby we would glue the windshield and rear glass into the car rather than having big rubber sections to mount the glass in. As engineers, we liked this process because what you were doing was taking the structural properties of the glass and bonding it into the body, thus increasing strength—whereas before glass was just a floating fixture in a piece of rubber. This made it easier to make smaller number one pillars, for instance, because we were using the strength of the glass.

Author: Stan Parker mentioned that Fisher Body had terrible problems producing the dies to make the car.
VK: Because of the crisp, sharp lines and surfaces, Fisher Body did have a lot of sheet metal tearing. I remember one of the die engineers showing us some of the panels they had run and how the metal was tearing. He asked that we add some surface in certain areas to keep the dies from tearing the metal. We'd take that information back and work it up on our clay models and work with the designers to solve the problems.

The first thing once notices about the 1969 Fleetwood Eldorado is that the headlights are not concealed. This was part of a complete restyling of the front-end, which included a finely textured grille. A new backup light was part of the fuel-filler door in the rear.

1970 Cadillac Fleetwood Eldorado, production model. The altered grille now had a heavier horizontal theme and the name "Eldorado" in script on the left side above a plate which read "8.2 Litre." This referred to the car's new 500ci 400hp V-8 engine. A protective molding was mounted horizontally along the center of the body and the name "Eldorado" was spelled out in block letters on the lower front fenders behind the front wheelwell. A powered sunroof was offered as an option for the first time since 1941. This was the last year for this first front-wheel drive Cadillac, but sales were at a substantial 23,842 units.

Author: So there was give and take on both sides. Both the designers and the engineers would make adjustments.

VK: That's correct. There was another aspect of this that was closely related. When a designer designed a car, he would smooth it out and make it beautiful. This would go to engineering and eventually you'd get to what we called "cube-up," where all the die models would be placed together (these die models were made of wood and had exaggerated curves so that the metal stamping would result in the proper final form).

All the dies would look lumpy. The reason for that was in the dies they would put in what they called "overbend."

The curve in the die was a bit more than specified in the released design, but would spring back to the shape called for in the released designed because of the characteristics of the metal. In other words, you had to "overstamp" it because the metal would spring back. So that's what we called "overbend." The die engineer had to know just how much to overbend the metal. It was an art in itself.

In later years, we could put all the information into a computer and it would tell us just how much to overbend a particular surface.

We'd have the same type of problem with glass. We'd design a piece of glass to have a certain shape, but glass was all gravity bent when it was made. A flat piece of glass is put on a framed mold that's run through an oven. By gravity, the glass just falls down and touches the mold. The guarantee that the glass is going to do that one hundred percent of the time is just not there. That's why you had different qualities of windshields.

Author: One automotive historian told me that the '67 Cadillac Eldorado could have only been built by General Motors because it was such an expensive body development project. Do you agree with that?
VK: There's no doubt, it was a very expensive project. As I said, part of our job was to obtain a realistic level of interchangability without affecting the individuality of each design. They wanted maximum design differential between the three cars, but with a maximum degree of interchangability. They wanted their cake and to eat it too. I was always very proud of the job the guys would do to meet those kinds of demands.

Author: Were you involved in any of the annual facelifts for the '67 design?
VK: Not too much. If there was a major reworking of body panels, yes, but normally facelifts meant the designers were working within given parameters. We got involved only if the change was major, such as changing a quarter panel or certain types of changes to a deck lid, for example. Such considerations as change in luggage capacity of the trunk or an effect on how a convertible top mechanism would work would call for our attention.

Author: You worked on an interesting convertible top mechanism back in the late fifties that was to have been used in Eldorados.
VK: Yes. That was the INFORA Top mechanism which would have solved several problems found in convertibles: First, the folded height, or the "stacking" of the top mechanism (when the top was down this mechanism could stick up about six inches), meaning that a boot had to be supplied as a cover. This was ugly, awkward, and the boot was often not used. Second, because of the way the top rails folded, a narrow rear seat was all that was possible for passengers in the rear of the car. Third, the mechanism encroached on the luggage space in the trunk of the car.

Del Probst, Ed Pedolan, and others in Advanced Engineering developed the Inward Folding Convertible Top Rail Mechanism, they called the INFORA Top which solved all those problems at one time. The idea was to have a flat surface between the back of the rear seat and the trunk lid, plus alleviate the other problems I mentioned.

Actually, they developed this originally with Corvette in mind, but to demonstrate the possibilities of the invention, they actually built a prototype 1960 Cadillac four-door convertible from a four-door hardtop sedan. We could envision this mechanism being used on Eldorados because of the luxurious space gained by using it.

We tried to sell the idea to Fisher Body, but ran up against the NIH Factor [Not Invented Here]. Even our construction of the four-door prototype did not change their minds.

The INFORA Top never got into production. The NIH Factor was just too much for it. Fisher would just keep finding a million and one reasons why they couldn't do it. Even when we showed it to them on our running car, the four-door Cadillac convertible. That was a completely functioning and tested car. We showed them they could build a four-door convertible without having rear doors that opened from the back (suicide doors) like they used on the '61 Lincoln. We had been very leery of hinging the rear doors from the number three pillars. That was a definite safety no-no and something that ladies did not like to deal with in getting in and out of cars. Doors were considered "vanity panels" for ladies back then when they were getting in and out of the car. Anyway, Fisher Body always found reasons to say "no."

Author: You worked on other Eldorados after the '67?
VK: Yes. I worked on the '71. We did the basic body structures in the Body Room. But the real problems came with the '79 and '80 programs.

People don't believe it, but that roof panel used on the '79 Eldorado and the bustleback Seville was the same identical roof panel.

Author: How is that possible?
VK: We did a restrike in the area of the drips (the terminal end of the roof gutter where the water finally runs out). The Eldorado had a "Devil's Horn" drip that was integral with the shape of the roof. The Seville just had a crisp break in the gutter. When we got back in the sail panel area, of course, we had two back light angles that the roof had to accommodate: a real stiff one for the Eldorado and a very slanted one for the Seville. Working closely with Fisher Body, at the point where the sail panels came up from the lower quarter panels to join the roof, we had to come up with compromised flange angles—welding angles, where the quarter panels were welded to the roof. Those two roofs came out of the same basic draw dies, but getting it all to work was one hell of an engineering challenge!

This was probably an even more complicated set of problems than was posed by the '67 Eldorado, because with the Seville involved, we were incorporating the development of a four-door car into the engineering. From a structural standpoint, this presented a whole new set of criteria.

But from an engineering standpoint, one of my favorite Eldorados is the '67 because it was a wild departure for Cadillac. It is still distinctive. You see one on the road and you say, "Hey, that's a '67 Eldorado!" You get into later years and you might recognize a car as being an Eldorado, but you won't be able to place the year. Sometimes people come up to me now and look at my '94 STS and say, "Gosh, that is a beautiful car, but what is it?" I love the car, it's probably the best automobile I've ever owned, but it's just not as distinctive as a '67. I've got an '80 Seville in the garage, which I'm gonna hang on to. I drive that out on the road and everyone knows what it is.

Anyway, I got a medical retirement from GM in '81 and didn't work on any of the later Eldorados.

One thing that I would like to add is that I think the "silent heroes" of Design Staff should be given credit for their work. The engineers and the modeling group are really the people that put the cars together. They're directed by a Chief Designer, but he didn't get his way a hundred percent of the time—maybe more like ten percent of the time. It was a hell of a team effort that went into putting any car program together.

It bothers me, as I've told you before, when I read that so-and-so did this car and so-and-so did that car. I think, 'yeah, but it took a hell of a team!'

That was one of the things I always admired about Bill Mitchell. He very, very seldom took credit for something, even though he might have been the instigator of it.

The Hooper bustleback Seville was a Bill Mitchell car from the word "go." He really pushed that design. Originally, he wanted it on an Eldorado. He lost out on that because Cadillac wanted the Eldorado to be a more stiff, sporty looking thing. But he finally got the design on the Seville. Mitchell always said, "My team of guys did the car." He gave credit to people like Wayne Kady who was Chief Designer at Cadillac Studio and had done the original drawings. That was Bill Mitchell: he was a tough guy, but he was fair. Authors' Note: Wayne Kady says that the bustleback seville was promoted by Bill Mitchell because Mitchell liked it. Kady adds that Kaptor eas unawake that Kady had done the bustleback about 1972 in 3/8 scale as a proposal for the 1975 1/2 Seville when he was Chief of Advanced Cadillac Studio.

Some people had a hard time dealing with Bill Mitchell, but I got along with him fine. You see, he didn't know my business. He'd see something in one of the design studios and he'd call me in. "Hey, Kap, I was just over in Cadillac Studio and saw such-and-such. Is that all right? Is it buildable?" I'd tell him something like, yeah, we might have to change it a little by doing a little wiggling here and there, but I don't think ten million Frenchmen could tell you the difference.

In fact, when we actually did the master models, we sometimes did some little wiggles and didn't tell anyone. Then we'd have the designers come in for the final blessing before we released it to the drafting room and they would bless it—never knowing the little wiggles we had to make to make the car buildable.

Chapter 5

1971–1985: Showboat and Flagship

The first front-wheel-drive Cadillac Eldorado had been a success in every way. The critics loved it, the public loved it, and automobile enthusiasts loved it. In fact, some enthusiasts love it so much that they refer to it erroneously as the "first" Cadillac Eldorado. Cadillac ads for newer models have featured it in the background, shrouded in mist like some chariot from Mount Olympus. The design is a classic.

Enter a Real Showboat

Following that act was no mean feat. The Eldorado that drove into the dealer showrooms for the 1971 model year was quite a departure from the first front-wheel drive version. It looked huge. It was the largest production front-wheel-drive automobile to this date ever built anywhere in the world! Because of some of the rounded lines, it even looked larger than it was. In spite of that, the car retained the personal nature of the previous Eldorado: two doors, sporty styling, long hood, short rear deck, and an impressive appearance. When you looked up and saw one of these behemoths in your rearview mirror, you knew there was something big on your tail!

If you look again at the design exercises for the twelve- and sixteen-cylinder project, you can see the natural derivation of the design for this car: the blade fenders, the pinched waist, the distinctive taillights, the massiveness of the fenders over the wheels, the pronounced front grille, and the long front hood are all there. This car was unmistakably Cadillac, but some considered it more showboat than flagship.

1971 Cadillac Fleetwood Eldorado, early scale model (April 1968). The totally new 1967s were still in the showrooms when this scale model was built. Notice the influence of the XP-840's almost separate blade fenders. The basic form of the 1971 is already there.

1971 Cadillac Fleetwood Eldorado, early scale model (August 1968). This is the same August 1968 model photographed from the rear. This angle shows off the slanted taillights and the almost fastback treatment of the rear end.

1971 Cadillac Fleetwood Eldorado, early scale model (August 1968). On the other side of the August 1968 model was a more conservative treatment closer to what was finally chosen for production.

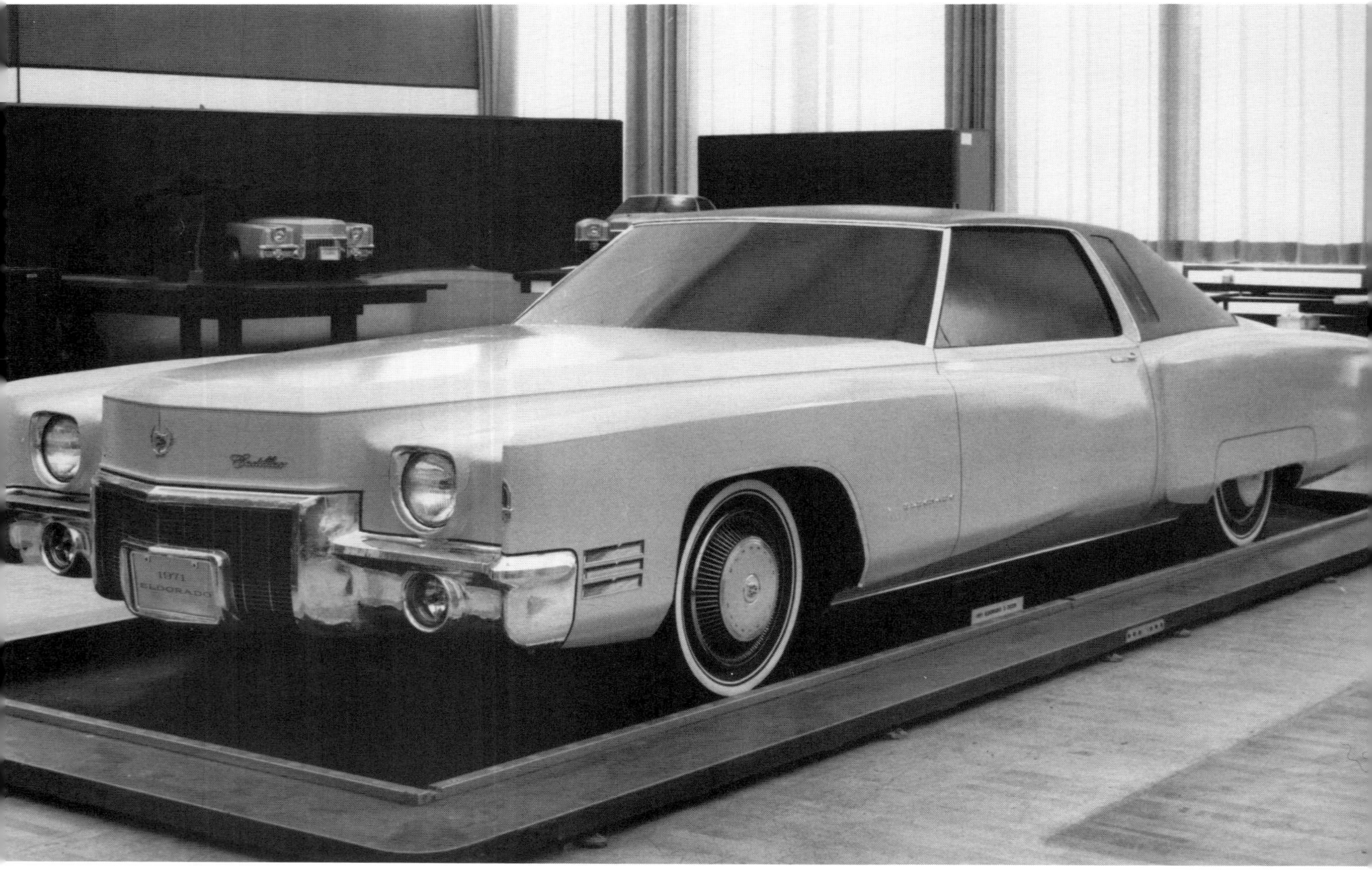

1971 Cadillac Fleetwood Eldorado, full-size clay model (September, 1968). The grille is not as radical as the August version, and there is an opera window in the sail panel. Somehow, the Cadillac script on the front of the hood doesn't work in the presence of the wreathed Cadillac crest in the center.

Again, the September model, but this time photographed from the passenger side of the car. Often designers will have a different design on each side of the car to save time and money. Here the right side of the car shows a four-door arrangement. Notice that the rear door would open from the front, what collectors call a "suicide door." There is no opera glass in the sail panel on this side of the car.

Retired GM designer, Pierre Ollier, remembers that the hood's outer panel was pressed from a single piece of sheet metal engineered by Wally Sitarsky. Ollier says Wally was the designer's dream die engineer; he always wanted to be challenged. He never would tell a designer that something couldn't be done.

Many people have said that there was a lot of Bill Mitchell in this particular Eldorado. Chrome was pretty much cut to a minimum, and the long, razor-edge crisp lines accentuated the body's shape. Indeed, Mitchell said the car would someday become a classic, and he kept one in his garage even after the introduction of the all-new 1979. The 1971 and 1972 versions of the car looked especially good. After the new bumper laws went into effect, Cadillac sort of lost control of its car's proportions. By 1976, some critics thought the car had almost become a caricature of itself. Some designers have said the body style was to Mitchell what the 1958s were to Harley Earl.

One thing you could say for Bill Mitchell, he had a feeling for what the American automobile buyer wanted. More importantly, he had the ability to know this years ahead of time, which was absolutely necessary considering the lead time needed to design and engineer a production car. This Eldorado satisfied a need in a group of luxury automobile buyers who wanted a vehicle that was personal, sporty, and would be a conspicuous emblem of their success. A car with this brazen flamboyance and size would fail in today's market, but the 1971 through 1978 showboats made money for GM.

The seventies were the years of flash—mini-skirts, bell-bottoms, and Elvis with pork-chop-size sideburns. Drawing attention to oneself was quite in fashion. Because of their immense size and the sheer acreage of the leather interior—not to mention the myriad gadgets and comfort appointments available—many consider these the most opulent and luxurious Eldorados ever produced.

Sales figures for this introductory year were reduced by a strike at GM. Only 27,368 Eldorados (20,568 coupes and 6,800 convertibles) were sold, as opposed to more than 40,000 in the following model year. In short, Cadillac had a winner.

The 1971 Cadillac Fleetwood Eldorado, the car's full name, actually was only a fraction of an inch (depending on how it was measured) longer than the previous year, but the wheelbase was increased by 6.3 inches to 126.3 inches (by contrast, the previous Eldorado had been 221 inches long, with a 120 inch wheelbase). A chromed nonfunctional vertical scoop was placed just aft of the door, emphasizing the pinched-waist design. The narrow vertical opera windows in the sail panel accented the formal vinyl roof in the coupe version, and the convertible version's Hideaway Top came with an optional hard boot that covered the retracted, almost flush, lowered top, as in earlier Eldorado convertibles.

Thanks largely to the persistence of Cadillac engineer Dan Adams, this was the first Eldorado convertible since 1966. Adams had been with the division since the 1930s and had always enjoyed and appreciated convertibles.

A sunroof and Astroroof (featuring a sliding tinted pan-

1971 Cadillac Fleetwood Eldorado, full-size clay model (November 1968). Sometimes when both sides of a model are not finished or one side is different from the other, a mirror is placed in the center of the model to get an idea of how the total design would look. Except for the grille extending to the leading edge of the fender and the wreathed Cadillac badge placement, this model is close to the final choice.

el) were available for the coupe. Curiously, the 1971 bodystyle went through several facelifts on its long run through 1978 without ever having concealed headlights. Hidden headlamps had appeared in the first two years of the previous model run and were a continuing feature with the Lincoln Mark III, IV, and V of the same period. Since the 1967 Eldorado's hidden headlight mechanism actually worked, and continues to be reliable with current collectors, the absence in the 1970s Eldorados was regrettable to many buyers.

The 1971 Eldorado coupe was almost $500 more than the previous year's. If you chose the convertible version, another $400 was required. Sometimes the sculptured front end is referred to as "coffin nosed," reminiscent of the 1930s Cord. If you say that about the front, you must say the same thing about the rear which is even more sculptured. A thin horizontal strip of chrome and wide chrome rocker panels protected the subtly rounded sides of the car. On top of the deck lid were the exhaust louvers for the Eldorados new flow-through ventilation. Automatic Level Control was part of the package in both coupe and convertible.

1971 Cadillac Fleetwood Eldorado, full-size clay model (December 1968). Here the line flow of the front fender doesn't reach the door handle as in the production car and there is no side scoop. The front bumper appears to extend to the leading edge of the front fender, and the sail panel is narrower than the final choice. Note the winged crest on the parking lights, which are mounted in the bumper.

This is the December model, a couple weeks later, from a rear three-quarter view. The rear-end treatment features short taillights and a decorative horizontal grille across the top of the rear bumper.

1971 Cadillac Fleetwood Eldorado, full-size clay model (December 1968). The day after the previous photograph, the model was moved outside to the turntable on the patio and compared with a 1970 Eldorado prototype and a Riviera boat-tail prototype. Note the triple horizontal cornering light arrangement on the front fender, the omission of the scoop on the rear fender, and the cut-out rear wheelwell.

Showboat Goes through Changes

For Cadillac's seventieth anniversary, the 1972 Eldorado received a few modest changes: a more pronounced grille frame, a vertical grille theme with "Cadillac" engraved in script on the left side, the word "Eldorado" just above the cornering lights on the front fenders, and the engine designation "8.2 Litre" engraved just before the door. Full wheel covers with concentric circles enhanced the appearance of the wheels.

Having recovered from the strike of the previous year, Cadillac produced 32,099 Fleetwood Eldorado coupes and 7,975 convertibles for the year. On three-day notice from then-President Richard Nixon, one Sable Black coupe accompanied him on his visit to Russia as a special gift to Soviet Union leader, Leonid Brezhnev.

Among the options for the model year was the Track Master anti-skid brake system for the rear wheels and the Custom Cabriolet roof, which included a specially framed sunroof on an elk-grained vinyl-covered top.

1971 Cadillac Fleetwood Eldorado, full-size clay model (December 1968). This is the same model showing a fender skirt on the rear wheelwell and one large cornering light on the front fender. Note the rise in the roof line as it approaches the windshield.

104

1971 Cadillac Fleetwood Eldorado, full-size clay model (January 1969). The following month, an interesting treatment that extends the rocker panel molding to the front and rear bumpers is tried on the left side of a model with a fender skirt and no scoop on the rear fender.

1971 Cadillac Fleetwood Eldorado, full-size clay model (January 1969). A rear three-quarter view showing the continuation of the rocker panel treatment with a scoop on the rear fender.

For 1973 the Fleetwood Eldorado sported a central prow on the front bumper under its new egg-crate grille. The grille itself was attached to the 5mph bumper for damage control. Distinctive parking lights wrapped around the leading edge of the front fenders, and the word "Eldorado" was moved to the immediate rear of the front wheels. Since the vertical fake chrome air scoop at the front of the rear fender was omitted, a long, thin strip of chrome could be placed on the side of the car running from the rear of the front wheel-well to a lighted wreath and crest insignia just before the rear bumper. This, and the presence of rear fender skirts, gave the Eldorado an even longer appearance than the year before.

The rear of the car was no longer sculptured, but appeared flat above a new straight, massive, 2-1/2mph rear bumper with central horizontal rubber strip. Of course, this necessitated a change in the design of the rear deck lid. The front hood had undergone a change in design also.

For the 57th Annual Indianapolis 500-mile race on May 28, 1973, a 1973 Cotillion White Fleetwood Eldorado convertible was selected as the Official Pace Car. Setting off this special Eldo was Indy 500 signage, custom pinstriping, and Goodyear tires with raised white lettering and a narrow white stripe. This Eldorado marked only the second time in the Brickyard's history that a front-wheel drive was used as the Official Indy Pace Car. The first was a Cord in 1930. It was the fifth time a Cadillac had supplied the Pace Car at Indy.

Another landmark for the history of Eldorado was that this was its twentieth birthday, but, alas, the marque's special year went without fanfare. Automotive historian Walter M.P. McCall has noted that after all those twenty years of inflation, the original (practically hand-built) 1953 Eldorado's price was

105

1971 Cadillac Fleetwood Eldorado Coupe and Convertible, production model. For the final production model, Cadillac chose to use rear fender skirts, a turn signal light separate from the front bumper and in the leading edge of the blade fender, a veined scoop at the front of the rear fender, no horizontal grille above the rear bumper, and a single horizontal cornering light on the front fender. Cadillac made the car available as a coupe and also offered the first Eldorado convertible since 1966. This car's wheelbase was 6.3in longer than the previous Eldorado, but less than 1in longer overall. A strike had a negative effect on production and sales were only 20,568.

1972 Cadillac Fleetwood Eldorado Coupe, production model. With GM's labor troubles over, sales passed the 40,000 mark, of which 7,975 were convertibles. This factory photo shows the new heavier vertical texture of the front grille, but the script above the cornering lights (a characteristic of this model year) is not present and could have been a late addition to the design.

1972 Cadillac Fleetwood Eldorado Convertible, production model. Here the script above the cornering light on the front fender is clearly visible. This beautifully maintained white convertible with red interior belongs to GM Designer Wayne Kady, shown standing beside the car. He was one of the primary designers of this Eldorado.

1972 Cadillac Fleetwood Eldorado Convertible, production model, rear three-quarter view.

still $390 higher than the price for the 1973 Eldorado.

The 1974 Eldorado had round quad headlamps and a fine-lined vertical grille, but most of the big changes were in the rear and interior. The new energy-absorbing rear taillight assemblies were connected by a reworked bumper, accenting a body that had been redesigned from the rear of the doors to the back of the car. A rear stabilizer bars was standard. Inside, new instruments were mounted in a restyled curved instrument panel. A limited number of driver and passenger airbag-equipped cars were produced during this model year.

The biggest change for 1975 was the appearance of redesigned rear quarter windows to replace the small verticals that had been characteristic of the car since its debut in 1971. Front parking and directional lights were moved and the Cadillac crest was placed on the side of the rear roof. The rear wheelwells were enlarged and the fender skirts omitted. The front grille was more square with an egg-crate pattern, flanked by all-new rectangular headlights.

Pollution control devices had greatly encumbered the 500ci V-8 engine, but for the first time fuel injection was available as an option.

Touted as "America's Favorite Luxury Car," the standard factory Eldorado price surpassed $10,000 for the first time. We remember a couple of TV gangsters of the time discussing a gambling debt in terms of the number of Eldorados the total would buy.

The big news for Cadillac in 1975 was the introduction of the new Seville. This car got people's attention and started talk of what was in store for Cadillac in the near future.

The "Last Convertible" Sideshow

The 1976 model year is remembered for the hype and clamor over the "Last Convertible." This was an unusual situation in the automotive world. Cadillac knew how many convertible top mechanisms were available from its supplier without the expense of placing a new order. Because of new pressures on the automotive manufacturers concerning automotive safety, the increased use of air conditioning, stereos, and other creature comforts available in coupes and sedans, and the percentage drop in sales of convertibles in general, the decision was made that this would be the last of the body type manufactured with Cadillac's name on it. The last Lincoln Continental convertible had been manufactured in 1966—a full decade before—and Chrysler Corporation didn't even produce its luxury Imperial models in any comparable form.

Money, or the lack of it, talks, and that's what automotive pundits say was the ultimate factor in the demise of the Caddy convertible: lack of sales. Sales of convertibles as a per-

1973 Cadillac Fleetwood Eldorado Convertible, Official Pace Car of the 1973 Indy 500. Cadillac issued the following news release on January 15, 1973: "This 1973 Cadillac Eldorado Convertible has been selected as the official pace car for the 57th annual Indianapolis 500-mile race May 28. Accenting this popular cotillion white convertible is an interior of red leather and a white top. Special striping on the beveled edges of the hood, sides, and deck lid highlight the subtle shapes of this classic front-wheel-drive motorcar. It is only the second time in the track's history that a front-wheel-drive car has been selected to pace the 500." (The previous time was a Cord in the 1930s.)

1973 Cadillac Fleetwood Eldorado Convertible, production car. Impact-absorbing bumpers were added fore and aft and the rear scoops were removed, making the car slab-sided with a rub strip running from the rear of the front wheelwell to the marker-light medallion forward of the end of the rear bumper. Restyling was so extensive that the car lost its coffin-nose rear end and the huge rear bumper necessitated flattening the rear end. Vertical taillights were built into the rear fender ends, where they have remained to this day. Overall, this model year was a taming of the original 1971 sculptured design.

centage of total Cadillac sales had dropped since a peak in the early 1960s. However, to be fair, the actual number of Cadillac convertibles hovered between 14,000 to 21,000 per year as long as both regular series Caddy convertibles and the Eldo convertibles were produced simultaneously. When the production of the DeVille series convertible ended with 1970, total sales of Cadillac convertibles took a nose dive, but, at the same time, GM had essentially taken away the less expensive convertible from the market. This made it easy for people to cite statistics that the convertible had fallen from a high of about 10 percent of total Cadillac sales to something less than 1 percent in the mid-1970s.

1974 Cadillac Fleetwood Eldorado Coupe, production car. The rear fender was redesigned, and the medallion near on the side near the rear bumper was omitted. The Cadillac name appeared in script on the left side of the header of the vertically textured grille. A new curved instrument panel graced the front interior.

1975 Cadillac Fleetwood Eldorado Convertible, production car. Now with an egg-crate grille and the Cadillac name in script on the end of the rear fender, this was the last American convertible still in production. Fender skirts were omitted and the wraparound front cornering lights were replaced by the horizontal rectangular lights. On the coupe, the narrow vertical coach or opera windows were replaced by large rear-quarter windows and the wreathed Cadillac crest appeared on the sail panels.

1976 Cadillac Fleetwood Eldorado Convertible displayed with sixty years of Cadillac convertibles. This illustration was part of Cadillac's announcement that the 1976 Eldorado would be the last convertible. Cadillac made the following news release on April 21, 1976: "Detroit—The Cadillac Motor Car Division of General Motors has been building convertibles for 60 years, and this past year Cadillac was the last American automaker to offer a convertible model. That great tradition ended today when Cadillac produced the last domestic production convertible on its Detroit assembly line. The 1976 Cadillac Fleetwood Eldorado is a far cry from the first Cadillac convertible, the Type 53 offered in 1916. Even as Cadillac convertible styling changed drastically during the past sixty years, the central theme of combined sportiness and luxury remained evident."

However, one should consider this interesting fact: In 1970, the last year of the DeVille convertible (there were no Eldo convertibles produced again until the following year), 15,172 of these ragtops were sold. That same year, Ford sold only 21,432 Mark IIIs, a car everyone considers a success and a car that certainly had many more parts specific only to itself than the convertible version of the two-door DeVille. Is that such a big difference in numbers as to warrant cancellation of a legend? It's difficult to understand. Does percentage of total sales tell the whole story when the reported figures almost reach that of a competitor's car that has been labeled "successful?"

In a December 1972 article in *Road & Track,* automotive writer and historian Allan Girdler considered many possible reasons for the decline of convertible sales in the total automotive manufacturing world. He discussed things ranging from safety regulations and national politics, to handling characteristics—as well as the fall in percentage of total sales. Girdler mentioned that, at the time, the convertible amounted to about 15 percent of Eldorado's sales, but he never considered the availability factor of less expensive models and the comparisons of total sales with other marques that were considered successful.

Edward C. Kennard, then Cadillac's General Manager and VP of GM, excused the decision to cease convertible production on the grounds of improved styling of coupes, increased use of automotive air conditioning, and a greater frequency of expressway travel at high speeds. He said that if 20,000 top mechanisms had been available, he could have sold every one of them that last year.

The mystique of the Caddy convertible had long been a part of American life. The first convertible had appeared on

1976 Cadillac Fleetwood Eldorado Convertible, a note on the last convertibles from Cadillac. Cadillac released the following comment to the news: "Detroit—The 1976 Cadillac Fleetwood Eldorado Convertible, the only American production convertible manufactured this year, has been classified by many auto enthusiasts as a collector's item even before its production had ended. Cadillac produced 14,000 of these highly sought after Eldorado convertibles during the 1976 model run, and every one of these front-wheel-drive convertibles was sold long before it was driven off the Detroit assembly line. Its companion model, the Eldorado coupe, will continue being produced." Cadillac's capacity for manufacturing the 1976 convertible was only limited by the number of convertible-top mechanisms the company could lay its hands on.

this country's roads as the 1916 Cadillac Type 53 Roadster. Cars with removable tops had been produced by Cadillac since Henry Martyn Leland started the company in 1902, but the 1916 roadster was the first that was actually called a convertible. Indeed, the first Eldorados of 1953 were available *only* as convertibles. This was a lineage that could not be abandoned without a great deal of soul searching and thought.

But, the decision was reached, the announcement made, and, what was called the last American production model convertible rolled off the Cadillac assembly line a little after 10am on Wednesday, April 21, 1976, a full sixty years after the introduction of the Type 53 Roadster. GM was inundated with requests by people who wanted to buy the very last production convertible. Cadillac took possession of that last car and placed a special Michigan Bicentennial license plate on it which spelled out "LAST."

However, the division produced a special edition of 200 all-white Fleetwood Eldorado convertibles, known as the Bicentennial models. These were identical replicas of the "last" car (the one Cadillac kept): white wheel disc inserts, white body with a white top, white leather seats trimmed in red piping, matching red carpeting, red and blue accent pinstripes on the body, and a red instrument panel. It was *the* car to drive to that year's Bicentennial celebrations. On the instrument panel in each car, a plaque proclaimed this to be one of the last 200 U.S. production convertibles.

The suggested retail sticker price for each of the Centennial models was $11,049. Speculation quickly drove the price up to $40,000 in some parts of the country. Automobile sections of newspapers were crowded with ads offering anywhere from $1,500 above purchase price on up for one of these convertibles in excellent shape.

The word "investment" was being thrown around by too many people who didn't understand the special-interest car market. Some naive buyers thought they could just buy one of the Centennial cars, use it as an everyday driver, and

1976 Cadillac Fleetwood Eldorado Convertible, the last one. The accompanying news release read: "Detroit—Pretty Ruth Ostyn, a purchasing employee at the Cadillac Motor Car Division, proudly displays the personalized license plate for a special Cadillac—the "last" domestic production convertible, a 1976 Cadillac Fleetwood Eldorado. The final U.S. convertible was manufactured on the Cadillac assembly line today and will be retained for historical purposes by Cadillac. This milestone automobile is registered in the state of Michigan and to add a further note of authenticity, the final convertible has been issued 1976 Michigan license plates which say it all in just four letters." Cadillac received an overwhelming number of requests from people who wanted to buy the very last convertible.

1977 Cadillac Eldorado Biarritz, production car. Although the name "Biarritz" had been used previously to denote the Eldorado Convertible on earlier models, now the name was being used as for a trim package on the Eldorado Coupe. The name was actually introduced late in the 1976 model year. Coach lamps on the sail panel and a special "Frenched" rear window were just two goodies in this special package. The Eldorado name was spelled out in block letters above the grille, which had now been reworked to a fine texture. The rest of the Cadillac line had been downsized for this year, leaving the Eldorado looking more like the huge luxury traveling machine that it was. For whatever reason, this was a record-breaking sales year for Eldorado—47,344 were sold.

sell it for a fortune at a later date. To have any chance at getting their money out of the car, they would have had to buy it and store it. This was an expensive proposition. Besides, production figures were huge, compared to other collectible cars that have increased in value substantially.

Automotive historian Richard M. Langworth reported that when convertibles were reintroduced about eight years later, a group of people who had bought Centennial model Eldorados as an investment attempted to sue General Motors. Their contention was that Cadillac had announced the 1976 was to be the last convertible and here they were producing new ones, thus destroying these "investors'" hope for future profits. Needless to say, this group did *not* win their case.

For 1976, 14,000 Eldorado convertibles were sold, 60 percent more than for 1975. Sales were limited by the availability of top-folding mechanisms. Two years before, every top mechanism they could lay their hands on was bought by Cadillac for use in 1976. Across the board, this was a good year for Cadillac Division with factories running at capacity for most of the year.

Eldorados for 1976 were featured standard four-wheel disc brakes, the word "Cadillac" in script left of the upper grille header, standard wheel covers with black-painted centers, massive bumper extensions at the bottoms of the leading edge of the front fenders, and a front grille with a vertical bar theme that extended higher than the quad headlights. The coupes had distinctive opera windows that sloped forward. The Custom Eldorado Biarritz option was introduced late in the model year as a special appointment package. This was the first use of the Biarritz name since 1964.

End of the Run

For Eldorado, the model year 1977 was like the day after the big party. For the first time there was no sexy convertible in the line-up, and frankly, the bodystyle was beginning to age. The big-bodied Cadillacs received a completely new and moderately downsized body that caused much discussion among buyers because it was presented as "The Next Generation of the Luxury Car." The public seemed to approve of the weight reduction and increase in fuel efficiency because sales of Cadillac's all-new full-size cars went up. Total Eldorado sales dropped from 49,184 in 1976 to 47,344 in the following year.

The body of the 1977 model was quickly identified by the appearance of rectangular side-marker lamps near the back of the rear fenders and the word "Eldorado" spelled out in block letters on the hood above the front grille. The extra-

1978 Cadillac Eldorado Custom Biarritz Classic, production car. This was the ultimate trim package for this body style's last production year. Special two-tone exterior paint and superb interior appointments made this car appear to be dripping in luxury. Despite the design's age, a total of 46,816 1978 Eldorados were sold.

cost Biarritz option was again offered with a special cabriolet elk-grained padded roof and, inside, Sierra-grained leather seats fashioned in a pillow design.

For the last year of its production, this eight-year-old bodystyle was produced in three different versions for 1978: the standard Fleetwood Eldorado (to which could be added the Custom Cabriolet package), the special-edition Eldorado Custom Biarritz, and, for this year only, the Eldorado Custom Biarritz Classic. This latter car received a special two-tone paint treatment and came with a special luxury interior. All 1978 editions are instantly recognizable by the heaviness of their cross-hatch-theme front grilles. Total sales were down from the previous year, but still at a respectable 46,816.

Classic Lines and Svelte Body

The completely new and moderately downsized 1979 Fleetwood Eldorado was big news among luxury automotive enthusiasts. The introduction of the 1975 Seville and the moderately downsized full-size Caddies in 1977 had prepared the public for this beautifully proportioned Eldorado. The car was powered by a choice of two engines shared with the Seville: the electronically fuel-injected 170hp 350ci V-8 or the 5.7 liter diesel. Four-wheel independent suspension and disc brakes made this an impressive automobile.

Actually, GM introduced three of these E-body cars: the Buick Riviera, the Oldsmobile Toronado, and the Cadillac Eldorado. We remember sitting in a restaurant with Ned Nickles who had recently retired as a designer for General Motors. He is best known for the 1963 Buick Riviera, the first Chevrolet Corvair, and the "port holes" on the front of postwar Buicks. We asked him what attracted the public to the Eldorado design over the other two E-body cars. Following him out to the parking lot, he pointed out the exquisite detailing in the design of his personal 1979 Eldorado. Once he called our attention to the contrast, it was impossible to miss.

The price of the new Eldo could get impressive fast. Starting at $14,668, one could add the usual array of accessories or opt for the Eldorado Biarritz package at $2,700 more with leather or $2,350 with cloth interior. An Astroroof further inflated the price tag. The Biarritz had a distinctive stainless steel front roof with a rear roof cabriolet treatment. Bright spears of metal rode the cap line to the base of the rear side light and forward to the leading edge of the front fender. Cadillac made a lot of money on this package. More people opted for just the cabriolet roof treatment, with the padding just rear of a painted front roof area.

This car was a natural for convertible conversion. In fact, some little conversion shops made a sloppy mess of cutting off the metal top and installing the convertible mechanism at outrageous prices. There were even instances of conversion without the required stiffening of the rest of the body to make up for the structural loss of the integral metal top.

1979 Cadillac Fleetwood Eldorado, clay model (February 1976). The other side of the same model shown in the previous photograph: a different side window treatment is tried, as well as an oversized front cornering light.

However, when done correctly, this Eldorado made a beautiful convertible.

For body identification by year, consider the following:

1980: New egg-crate grille and slotted wheel covers.

1981: Standard wheel covers had large red centers, and the front grille was fine cross-hatched.

1982: The Eldo had black bumper rub strips and three horizontal crossbars on a grille with a vertical theme and the Cadillac crest on rear lights. Also, the Touring Coupe (instantly recognizable by its hood ornament which is flush with the hood) was first offered midyear and a "Full Cabriolet Roof" option was made available, which was actually a simulated convertible top available in black, white, or dark blue.

1983: The word "Cadillac" in script on the driver's side of the front grille and aluminum wheel options were made available for the regular Eldo.

1984: Body-colored side moldings and rectangular headlamps that sat over clear parking/signal lights were featured. A Cadillac Division-sanctioned convertible (manufactured correctly by American Sunroof Corporation, Inc.) was now available for a whopping $10,944 premium over the price of the regular coupe.

1985: The front cornering lights held an orange square at their leading edge and two narrow horizontal reflectors appeared in the rear bumper. It was the last year for this Eldorado convertible conversion, but without the fanfare of "The Last Convertible" in 1976.

Engine Changes

For 1981, an unusual engine system was introduced which caused much unhappiness among Cadillac buyers. The 1979 Eldo had begun with the 350ci V-8, replaced by the 6.0 liter in 1980 with digital electronic fuel injection (DEFI). For 1981, Cadillac was at a loss about what to do.

The CAFE (Corporate Average Fuel Economy) standards were closing in and the cast-iron block, aluminum head HT4100 engine (HT stood for High Technology) was not ready for production. An interim measure was needed.

Thus, the V-8-6-4 was introduced for 1981. In theory this was a have-your-cake-and-eat-it-too engine. An electro-mechanical control system was supposed to shut down two or four of the eight cylinders as the power requirements changed. The idea was that this would save fuel when it was not needed. The word from the engineers at Cadillac is that the basic idea of the system worked, but there were pesky electrical problems that resulted in customer dissatisfaction.

Overview

The showboats of the early 1970s were, above all, big beautiful machines, well balanced, and comfortable to ride in. In many people's minds, the elegant, classic lines of the 1979 styling represented what a modern Eldorado *was* or *should be*. The genius of designer Wayne Kady was that he could oversee the creation of such lasting tributes to the spirit of Eldorado.

Wayne Kady Interview

1967–70 Eldorado, 1971 Eldorado, 1986 Eldorado, and The Twelve- and Sixteen-Cylinder Cadillac Design Projects

Wayne Kady graduated from the Art Center School of Design in Pasadena and joined GM in February 1961. After a short stint in design development, he went to Cadillac Studio. Kady has had more direct experience in designing Cadillacs than any other designer interviewed for this book. He has worked on the designs for several Eldorados, including the primary models of '67, '71, '79, and '86, as well as many of the subsequent facelifts. Kady was chief designer at Cadillac Studio from 1974 through 1986.

Author: What led up to your first experience with Cadillac design?
WK: At that time I was working for Bernie Smith in the Preliminary Design Studio for about six months, maybe less than that, and we were working the themes for the '65 Cadillac program along with the Cadillac Studio. As it turned out, the design that was in preliminary design was favored over the one being done in Cadillac Studio, and I was transferred to Cadillac Studio with the car. So my first involvement with Cadillac was on the '65 program. I was a young designer and this was my first experience with a production car. I also worked on the '64 New York World's Fair show cars, but the production '65 Cadillac was my first production involvement. For '65, we took the fins off.

Author: I love that '65 Eldorado convertible. It's beautiful.
WK: A friend of mine has one. That was a rare car. There were only 2,125 built.

Author: Of course, there wasn't that much difference between it and the other Cadillac convertible of that year.
WK: No, it's mostly interior differences. The Eldorado had just a slight trim difference on the exterior. But other than that, it's the same car.

Author: You worked on several Cadillac Eldorados, besides the 1967?
WK: Yes. I worked on the '67 through '70 and then was responsible for the '71 Eldorado. That car was designed end released for production in what was then called Advanced Cadillac Studio. I was the chief designer for the '71 Eldorado. Later, I worked on the '79 and then the '86 while in charge of the Cadillac Production Exterior Studio.

Author: Some say you were responsible for the '67.

Wayne Kady.

WK: No, I wasn't. The guy who was in charge of the studio at the time was Stan Parker. He was the chief designer of the Cadillac Studio when that car was released for production. There were others that were involved before him, especially Chuck Jordan. That car was in development for quite a few years, but the final design was under the direction of Stan Parker.

Author: He is retired now?
WK: Yes. He retired a number of years ago and he's still living in the area. In fact, I just saw him last summer. Because I own a restored '67 Eldorado and happened to be in his neighborhood, I thought it would be fun to stop in on him. When he came out to the driveway to greet my wife Marie, and check out my car, he was thrilled and talked of fond memories of that design program. We had a nice visit.

Author: Were you involved in the twelve- and sixteen-cylinder project?
WK: Yes, but my involvement never went anywhere beyond sketches and a few renderings.

Author: How long did you guys spend on that project?
WK: On the sixteens? I don't know. It was just a wild dream. It was a filler program in between major programs, a fun exercise in between the bread and butter stuff. I don't remember spending all that much time on it—maybe six months more or less. It might have been something like that. The project was mainly sketches and I think there were a couple of models built; that was about as far as it went.

Author: Were you in preliminary design then? Was that project the beginning of the '67?
WK: No. I was working in the Cadillac Studio at the time. That was when it was under the direction of Stan Parker. The project started out as XP-727. That was its code name, and then it evolved through several themes before it settled into what became the production version.

Author: Well, the reason I'm asking about that is that Dave Holls had said that the car was under development in one of the experimental studios or something before it got to Cadillac.
WK: Yes, we were working in a walled-off part of the preliminary design studio. But it was a completely different car then, from the way it ended up.

1979 Cadillac Fleetwood Eldorado, clay model (February 1976). At one point, designer Wayne Kady says, the "bustle-backed" body that became the 1980 Cadillac Seville was considered to be the replacement for the 1971-style Eldorado. The model in this picture was created as a possibility when the designers were working toward what became the 1979 Eldorado. The rear end's similarity to the bustle-back 1980 Seville is readily apparent.

Author: Oh, it was?

WK: Yes, it was code named XP-727 while it was in the preliminary design studio. It then featured a severe vee'd front plan view, egg-crate grille flanked with large round headlamps mounted midpoint of the front fenders, the side had round flared wheel openings, and a grille mounted behind the front wheels. The rear had thin vertical lamps mounted on the trailing edge of the quarter panel. It was a dramatic, classic looking design.

Author: Well, I noticed a lot of similarities between the production '67 and the sixteen with the blade fenders. But I'm not a designer. Are you just agreeing to be nice, or is that the case? I don't know.

WK: Well, there are completely different proportions in the '67 compared to the sixteen that you're talking about, but trying to remember exactly how the design evolved is difficult. But I'm sure that in that program we covered many directions and there were influences and cross-pollinations going on. The best of all that we covered ended up in production.

Bill Mitchell.

Author: Okay, but it seemed to me that those blade fenders became a characteristic of Eldorado. They were on the '67 and appeared all the way through the '79 in an altered form. Is that something you guys wanted to be kind of a signature of Cadillac?

WK: Yes. We talk about a brand character today and we're trying to make a lot out of brand character with all our car lines. We were dealing with brand character back then. Brand character is design cues that are owned by that brand. Some of the design cues for the Cadillac were the blade front fenders and the rear quarters with the vertical taillights, and, of course, the egg-crate grille.

Author: And you still have the horizontal grille that's been a part of Cadillac since '41?

WK: Well, yes, that's right. It goes back to about '41.

Author: You took over Cadillac Studio to do the '71?
WK: Yes. The '71 Eldorado was designed in the Advanced Cadillac Studio. That was my first responsibility as a chief designer.

Author: Back in 1979, I was at Bill Mitchell's house and he had a '71 Eldorado . . .
WK: *Blue Boy.*

Author: Yes, that same design. Oh, that's what he called it?
WK: Originally Bill had that car painted gold with all the chrome parts, including the bumpers, gold plated. After seeing the movie The Great Gatsby, he was convinced that it could be a new trend—the pastel colors, the wide whitewalls, the aura of that period. So he had the chrome parts restored and the car repainted a light blue nonmetallic color with a lot of white accent striping, and called it *Blue Boy.*

Author: Did Mitchell have much input into the project?
WK: Of course, you know, he was the boss. He's the

1979 Cadillac Fleetwood Eldorado, full-size rendering. Late in the game, a design variation is compared with a photograph of a clay model.

1979 Cadillac Fleetwood Eldorado, full-size rendering. Here a two-door variation of the 1970s Seville is attempted to determine suitability as a replacement for the Eldorado.

guy that picked the designs and he was the guy that made sure they got sold through the hierarchy of the corporation. As far as him designing the cars, I wouldn't say that he designed them. He would OK the designs. He was vice president and the director of Design. He always got credit for everything. And rightfully so. He was the boss.

Author: Well, he had to get things through, but coming up with the art is a different matter.
WK: I was involved in the '80 Seville too, and that was a hard sell and if it wasn't for Bill Mitchell, we wouldn't have sold that design. He pushed that through with Pete Estes. Originally that design was one proposed for the '79 Eldorado, but Ed Kennard, the Cadillac general manager and vice president, didn't like it. We had experimented with that design for the '76 Seville.

Author: Didn't you do the Seville before that?
WK: We had scale models going back to the 1975 Seville. We had bustle-back designs that were very similar to that.

Author: No, I mean the model Seville before that. The '75 Seville—the one that's so popular now. Were you involved in that one?
WK: Yes, but only in the early stages in Advanced Cadillac Studio. The project was taken over by Stan Parker and I moved to Buick Studio. Stan Parker was responsible for the '75. I worked on the '80 Seville, the bustle back.

Author: What was your involvement in the '79 Eldorado?
WK: I was completely involved from conception to production. We reduced the body size considerably. We brought it down from about 224 inches to 204 inches.

Author: Is that the size Cadillac wanted?
WK: Yes. After the '76 Seville came out, we saw that our standard line of cars was just getting too big. Wasn't 1973 the first big energy crisis?

Author: Yes.
WK: Fuel economy was one of the big problems and the only way to get fuel economy up was to downsize the cars and get the mass out of them, and that's the reason behind the Seville and the downsized '79 Eldorado. By then the Eldorado was longer than the Brougham and heavier than the formal limousine. It was just too big and fuel thirsty, antisocial looking.

Author: Sometimes I know you guys do the most expensive car first and sometimes you do the least expensive

Drew Hare was Chief Designer for Cadillac Design Studio, Interior, for several years. He was responsible for many of GM's finest interiors. Hare worked with George Moon in developing the interior design for the spectacular 1963 Buick Riviera. He also did the interiors for the 1979- and the 1986-style Eldorados. Hare is now retired from GM.

one. I was trying to understand when that stainless steel roof car came out.
WK: 1979.

Author: Did you design it first and the other ones came out of it, or was it the other way around?
WK: Well, it was the other way around. We designed the base car first and then the stainless steel was added to it. In fact, the Biarritz was a moneymaker for Cadillac. They made a lot of money off that version—moldings and a stainless steel cap. And then the interior, of course; it had a different tufted leather seat design. By 1983, almost one-third of Eldorado production was sold with the Biarritz option.

Author: Who did the interior on that one?
WK: It was Drew Hare, who is retired and living in California.

Author: What was the basic idea behind the '79 then?
WK: The basic idea was to maintain the flavor of the '67 in a reduced size. It still had the blade fenders. It still had the vertical taillights. It had a lot of the flavor of the '67 except in a smaller, more contemporary size and proportions. It was a little less complicated because of its size. It still had the formal roof line and vertical D-line. And, actually, it was a pretty efficient package.

Author: The trouble with me is I get in a car and my head hits the top too often. That's what happened in the '79 Eldorado back seat. The next Eldorado got even smaller.
WK: Yes, that was a problem with the '86 Eldorado. Cadillac wanted that car downsized to the point that if we were to err, that we err on the small side. Fuel economy was the main concern and Cadillac was concerned with maintaining their dealerships. The price of fuel was thought to skyrocket by 1986 and there was also the threat of rationing. Originally the car was to be released seventy millimeters narrower, and at the last minute the drawings were separated at the centerline and widened to accommodate a V-8. Originally, a V-6 was the program. Had we stayed with that size, it would have been a total disaster. It just didn't look like a Cadillac, and sales bore that out. With 20-20 hindsight, the gamble to design for the energy crisis was lost.

Author: You've got such a long lead time between design and production. It's almost like you have to foretell the future. I'm sure you agree that the '86 is the most disappointing Eldorado?
WK: Yes—by far. That one I'd like to forget.

Stan Wilen Interview
Bill Mitchell and Eldorados of '71 through '92

Stan Wilen graduated from Pratt Institute in Brooklyn, New York, in 1953, then started at GM. After a short orientation period, he worked briefly with Buick, then served in the military between 1954 and 1956. Wilen's first assignment after service was in the Pontiac Studio. He started as chief designer for Cadillac in February 1968 and stayed until September 1974. Wilen retired from GM in 1991.

Author: What was the first Eldorado you worked on?
SW: Well, actually the first Cadillacs that I worked on were the '70 Coupe DeVille and Sedan DeVilles. We just had to take care of the finishing details. Then I think Len Casillo worked with me on the '73 Eldorado which was already under way. Len was assistant chief designer.

Author: One afternoon back in '79, Bill Mitchell invited me over to his house where he and I went into his garage. He had driven over to his house from the GM Tech Center in his Camaro with the Ferrari engine in it. In his garage was this Eldorado convertible. I don't remember if it was a '71 Eldorado there, specifically, but it was a blue Eldorado of the '71 bodystyle. Wayne Kady told me a couple of weeks ago that Mitchell called that car *Blue Boy*. Anyway, he walked over to that car and put his hand on it and said, "Someday this is going to be a classic."
SW: He did? Bill Mitchell was nothing if he wasn't enthusiastic! He might have been overconfident, too. Somehow when you mix enthusiasm and overconfidence you get a mixture that was the substance of Bill Mitchell—in addition to all the other things that he was. I don't mean that to be a negative comment; he was a little boy 'til his last breath, especially with regard to cars!

Author: He had to be a showman to a certain degree, because he had to sell that stuff to GM management.
SW: Yes, he was. Let me tell you a story. I went to one of the classic car shows at Meadowbrook. It was a hot day, really hot air. Bill Mitchell was walking around in a white suit, with a white hat, he had a white vest, white shoes, white socks, and a white shirt with a peach tie. He looked absolutely beautiful! He had driven up on his motorcycle and he was walking through there, very uncomfortable that neither he nor his motorcycle were drawing any attention. He was moving through the crowd trying to get attention. That was Bill, he wanted to be at the center of attention. If his cars weren't at the center, then he wanted to be there. Bill was a showman, but he usually had good stuff to sell.

Stan Wilen.

Author: So, when you first joined Cadillac, the '73 was already under way.
SW: Yes, I remember that when I went into the studio it was already sitting on the platform. We worked on the wheels. I then accepted the programs for '74 and '75. We did some grilles, over twenty, as I recall, before we released one. I know we worked for six months on a grille header for '75, some other front-end details, and some front fender changes. We also changed the shape of the rear quarter under the quarter window. Those were subtle changes on the fenders and quarters. The intention here was to advance the design, not start over.

Author: Were you there for the next Eldorado, the 1979?
SW: No, I don't think I actually started any of the Eldorados when I was chief designer at Cadillac Studio. Wayne Kady was my assistant, before he was promoted to chief at Buick, and George Moon was in charge of interiors. George was by far the best interior designer at GM. George was logical and had marvelous taste. He was almost always right on anything he ever did. Because of his expertise, he became head of all interior design at GM and he was succeeded at Cadillac by Drew Hare, who was followed by Marv Fisher.

All the chiefs of the various studios worked for Irv Rybicki, and his assistant was Jack Humbert. Those were difficult days to do an Eldorado. The '67 Eldorado had been a hard act to follow. Our changes on the '71 showed that.

I remember there were some conflicts with George Elges, [who was General Manager of Cadillac, and Design Staff]. Some of the people at Cadillac were sending the cars out to the aftermarket fellows like Cars & Concepts and American Sunroof. These companies were putting sunroofs, half vinyl tops, and targa tops on these cars. Some of the things that were being done were terrible: for instance, one company had put gold bars on the sail panels. Changes to the Eldorado coupe, especially, created a lot of problems. The people at Design Staff were concerned that their function was being compromised. Here

Above and below, 1979 Cadillac Fleetwood Eldorado, full-size clay and fiberglass model development.

they had designed the cars to be elegant and the division was taking it upon themselves to make modifications to the designs.

Author: Do you remember any specific events related to Eldorado design and these aftermarket vendors?
SW: I remember that one day we were out on the patio with Bob Templin (the chief engineer for Cadillac) and George Elges. We were proposing a taillamp for use on the Eldorado. Bob Templin was strongly against it, mainly because it was late, and, since he was against it, Elges was sympathetic for it. They had that kind of relationship. We at Design Staff were pretty sensitive to these sorts of things and were not above exploiting them.

Templin was a very fundamental person. He was a bright engineer, but he just didn't have that extra dimension. You could show him the most beautiful car in the world and, if it was twenty minutes late, he would reject it.

Anyway, after a few minutes of discussion, Elges bought the taillamp and then Templin gave a short speech on lateness. Templin just wanted to get at least something out of the conflict.

I stood there and listened to what Templin had to

say. At first, I decided I wasn't going to fight back, but then, I guess, he fed me too big a dose.

I said to him, "Templin, do you want to know why we're late? I'll tell you why we're late. We're late because I'm driving an 8-6-4 Eldorado and every day that I come into work I've got to get towed the last four or five miles! That's the reason I'm late! What do you think of that? Have you got any more questions?"

Elges laughed so hard, I thought he would break something. Everyone knew Templin had been responsible for that engine. With tears of gratitude in his eyes, Ted Hopkins, the Cadillac sales manager—the guy who had had to explain the shortcomings of the 8-6-4 engine to the dealers—embraced me physically there on the patio. From then on, the relationship between Bob Templin and myself was adversarial. That's the kindest way to put it.

1979 Cadillac Fleetwood Eldorado Biarritz and 1957 Classic Brougham in comparison photograph. This picture was released by Cadillac to the news media on September 20, 1978, with the following caption: "1979 Eldorado Biarritz is distinguished by stainless steel roof cap, first on an Eldorado since the Classic Brougham in 1957. That car was priced at $13,074. Adjusted for inflation, it would cost $29,979 today." The Biarritz option, which included a stainless steel front portion of the roof was offered as an option when this model Eldorado was first introduced.

The new 1979 Biarritz with the brushed stainless steel roof and cast-aluminum wheels. All Eldorados featured new independent rear suspension and electronic level control.

1980 Cadillac Fleetwood Eldorado, production car. This year the car featured a bold cross-hatched grille with dominant vertical bars. The upper part of the grille bore the word Cadillac in script. Electronic fuel injection increased fuel efficiency.

1981 Cadillac Fleetwood Eldorado, production car. A delicate crosshatched grille was centered between the quad headlights. Large, red-center medallions distinguish this car from the previous year's. The interior had a new center console and a greater use of wood appliqué trim.

1982 Cadillac Fleetwood Eldorado Biarritz, production car. For this year, the grille displayed a vertical motif complemented by three horizontal bars. Front and rear bumpers supported black rub strips with white centers. The Cadillac crest insignia decorated the taillights.

It was after that confrontation that we had our biggest run-in with Cadillac Division over their unauthorized use of outside vendors.

We had gone down to Cadillac on a problem and we had seen an Eldorado parked in the garage. It had an awful aftermarket padded roof on it. I asked my guys to find out where this car came from. We found out through Fisher Body that the division was releasing it for production.

It was my responsibility as chief designer at Cadillac Studio to report this to Bill Mitchell. I went to him and told him that not only is the roof ugly, but I don't want you hanging me from a tree because you're going to see this on a Cadillac—but I don't have the steam to tell a vice president (Elges) to cut it out. Since we didn't know anything about the car and we had to go to some difficulty to find out where it came from, it appeared to me that Elges was trying to circumvent the release process at GM. That kind of thing was supposed to come before a policy group and there were supposed to be checks and balances. The system required that such things be presented in open forum. Besides, what else was slipping through?

Mitchell said, "Put that on paper." I did.

That afternoon there was an EPG (Engineering Policy Group) meeting and Ed Cole, the president of GM, was chairing it. During the meeting, Mitchell slides this note that I had written to him, over to Cole. This was the wrong thing to do because I had just put the facts on a piece of paper. If I had known that it was going to the president of the company or to Elges, my language would have been different! I would not have been so accusatory. Cole read the note and passed it over to Elges.

Elges got red, then got white and left the room. I spent the next five months trying, unsuccessfully, to communicate with Cadillac Division, but they had gotten very strict instructions to have absolutely nothing to do with "Design Staff, Wilen, Mitchell, or any of those jerks down there!"

This confrontation really began when the vendors realized there was no sense going down to Design Staff because they rejected almost everything. Most of the time, we rejected things like that because they were ugly. But the fact was that Design Staff wanted to do Cadillacs and Buicks and Chevrolets and all the other GM cars and trucks—we didn't want some outsider of questionable judgment doing them!

But at Cadillac Division, where they were having some problems because of the energy crunch, said, "Look, it's our business to sell cars and, if Design Staff doesn't like it, we're sorry about that but we're going to sell cars." The philosophy was that when the dealer buys the car, he owns the car. If he wants to put junk on it, that's up to him, and Design Staff can go to hell.

So the vendors started bringing all this stuff to the divisional sales and marketing managers, who were more receptive. Suddenly, the worst stuff you ever saw began to show up in advertising for local dealers. Corporate executives, including the president, would ask us about them. We were in the dark, but we were responsible.

Design Staff protected the taste of Cadillac vigorously.

Mitchell really hated stuff like that and would call these add-ons all kinds of things. I remember him calling an aftermarket vinyl top "pig bladder." Bill would use the most demeaning terms. The trouble was that this only stiffened the backs of the division heads, because they didn't want their taste put down in front of their people.

Finally, Design Staff was put in as judge and jury on all matters of design and taste. Actually, it should be that way, that's what we're paid for. We are the ones who're going to get blamed for something if it's not right. But that degree of authority was resented by some—by many!

Author: What do you think is the future of Eldorado? Do you think it will remain the flagship of Cadillac?

SW: Normally, I would say yes. It is the connection with whatever you'd call the "young Cadillac customer." Now that I've been retired for a few years, I look around and see that if people are young—either in attitude or years—if they want a Cadillac, that's the one they'll get.

But these aren't coupe years. I remember a time when all you had to do was produce a coupe and people would buy it. Coupes come and go in cycles. Sedans now are better proportioned. Everybody is coming up with the same phenomenon of the sports sedan.

The best years of my career at GM were at Cadillac. That was because the aesthetic goal of the Cadillac was the clearest in the corporation.

Anytime you had a design decision, you settled it by deciding which was the most elegant—which choice had the most refined taste. That concept was at the bottom of all decisions. I could go to Oldsmobile, or to Pontiac, and you'd find that in some cars they'd want to sell excitement and in some cars they'd want to sell class and in some cars they'd want to sell economy and in some cars they'd want to sell youth—but not Cadillac! Cadillac went for a mature, accomplished individual who wanted elegance and taste. As a designer, I could reach with absolute confidence and clarity for that quality. I could also defend my decisions that way.

I remember working on the bumper ends of the '73 and '74 when those government rules started coming in. We wanted to have a vertical accent in the front because no one was doing that, so we had these blade fenders with a vertical character—this was on the '75. We had to allow for the bumper to move back into the body. The vertical thrust became the rocker molding. It was like an undercarriage under the whole car coming back into the taillamps. It was a very nice theme, but it was so cut up because of the government bumper rules and the fact that the bumper had to be able to move back into the body, and the rocker moldings didn't (they had to stay fixed). The idea was good. The execution was hard to work out. We might have done a better job ten years earlier, before there were such rules.

The multiplication of rules during that time made it harder to do elegant stuff. These big, beefy bumpers—that keep trucks, streetcars, and low-flying planes out—made it difficult. I think that now we have learned to handle such things better. But not back then.

1983 Cadillac Fleetwood Eldorado, production car compared to the first front-wheel-drive Eldorado. Note that the 1983 is wearing wire wheels and no vinyl roof.

Author: What do you think about the current Eldorado?
SW: I worked on that Eldorado. Dick Ruzzin was the chief designer. I played a very, very trivial role.

One day, a Saturday, we were all in the building. We were late on everything. Dick Ruzzin reported to Dave Holls and Len Casillo at the time. I was just leaving one of the Chevrolet studios and was heading for another one when Dave yelled out at me. He said, "Stan, come here, I want to show you something." This kind of thing happened all the time. Sometimes you'd just want to get another view from someone who was absolutely unattached to a project so you would get a fresh perspective on it.

So I went into the Cadillac Studio. I could see it immediately. They were being seduced into this trend to soft design. Trying to stay current and up-to-date, they were making a compromised version of it. They knew an Eldorado had to have dignity, so you don't make it too round. On the other hand, to make it crisp or harsh would have been behind the current trend. So I put a couple of pieces of tape on the model to show a D-line and a backlight rake on the side view. It was a classic Eldorado.

Dave said, "Now that looks like an Eldorado, but how are we going to sell that to Jordan?" I said, "Look, I've got my own problems. You guys are going to have to sell it to Jordan. I worked on Eldorados for six years and I'll tell you what an Eldorado isn't: Every time you put a round line on an Eldorado—unless it is a very easy, leaping front fender coming out of the door—you just don't have the elegance or the important look that an Eldorado needs. It has to have dash, but it's got to also look important, or it's just another car."

So they made a mockup over the rest of the weekend. Now I wasn't there when Jordan saw it, I was just told about him seeing it. "Well, I guess it's true. The Eldorado has to be harsh like that. It has to be chiseled, and have a jabbing line," Jordan said. Cadillac Division came in and saw these changes and liked them and they continued on with the design work. So I played a small role in developing that Eldorado—a very minor role, I acknowledge this. I'm just very pleased that I was able to do that.

Author: So you were addressing the philosophy of what an Eldorado is?
SW: Yes, there are certain things you can do with an Eldorado. You can put character into the surfacing: the wash of the surface as it comes from the top profile into the body and down the rocker. You have some choices there, but the hood has to stand proud. The rear profile and the front profile, the taillights and the headlamps have to have a decisive, stately look. It has to say, "Nothing wishy-washy here!" Rolls-Royce does this. You can put a flowing line through the center of the car if you need to. The '73 did, for instance, but we had a bevel, and below it, a harsh line that went horizontally through the car. If you look at a '67 Eldorado on through the ones on the road today, especially, you'll see these words hold true. You can see vestiges of those words in all of the Eldorados, including the current one.

Henry Meyer Interview
The 1985, 1986, and 1992 Eldorado Programs and the Development of the Northstar System

Henry Meyer joined GM and began studying at GM Institute in 1955 sponsored by and working in Cadillac. He spent his entire career in that division. Meyer was chief engineer for the 1992 Seville/ Eldorado project.

Author: Someone told me that at some meeting you were in during the early eighties, a three-cylinder Cadillac was actually considered.

HM: My recollection of that is that at the time—and we were talking about cars such as the '86 series of cars the size of Eldorado/Riviera/Toronado—the actual proposal was from Buick. We were all afraid that by that time gasoline was going to be three dollars a gallon and everybody was thinking in terms of smaller package sizes. We never really seriously considered the three-cylinder for a Cadillac, but we did consider V-6s for Cadillacs at that time. [A 4.1 L V-6 was offered as an option with several Cadillac models in 1981, including the Eldorado and Seville.] The only V-6 Cadillacs that were made were the Cimarrons. I can't remember the years for that. That was about '75, during the first real fuel scare for a very short time. That's about the time GM tried the diesel engines. This kind of solved the fuel economy problem. But, to my knowledge, the three-cylinder was never really considered for the Cadillac.

Henry Meyer.

Author: Were you involved in the 8-6-4 program?

HM: Not directly. At the time I was a body engineer. But as I recall, there really wasn't a problem with the 4-6-8 concept itself. The problem was electrical, but any problem with an engine will tend to give the entire engine a bad name. It was an ingenious concept, though. I do remember going to dealers for clinics for the '86 EK models (Eldorado and Seville). I remember one in particular in Houston in the Galleria area in 1981. I talked to the dealers there, gathered some information, and took the concerns that they had to the company. But I wasn't a powertrain engineer at that time and was only tangentially related to the program.

At that time, the European cars were very strong and we did clinics at which we did evaluations of Mercedes and BMW versus the Eldorado, Seville, Riviera, and Toronado. We got an idea of what the customers thought. This was one of our early attempts at getting an idea of what the customer wanted. At that time, I am not sure we were able to interpret our findings correctly. By the time of the '92 program, things certainly improved a great deal.

Author: How did the '86 program go over when you did the clinic you mentioned?

HM: We didn't show them styling at that point. In fact, the styling didn't exist. We were just getting some early feedback. We had a few renderings that they looked at, but as we sought the voice of the customer then, as I said, we didn't know how to interpret their responses. What the Cadillac customers of that time said—and you have to remember that many of them were driving very large cars at the time—was that the cars in the renderings looked too small, that Cadillac means big. But we thought everybody was going to be downsizing because we thought a big fuel shortage was coming.

As a result, the customers just wanted what they had, which were very large cars. We felt that they didn't understand that this fuel crisis was going to come—which it didn't, as it turned out—and that they didn't understand the oncoming technology. While we listened to what they had to say, we put too much stock in the "crystal ball forecast" of coming fuel economy difficulties.

In the end, the voice of the customer was pretty accurate because they did not like the size of our cars when we finally brought them out in '86. As you know, that program was not as successful as earlier ones.

In '85 the DeVille Fleetwood program was initially not well received either—because of its downsizing. While the interior creature comforts were just as good and better as previous years, and many people enjoyed driving them, when they left the car to walk into the house and looked back at the car, they said, "Gosh, is that really a Cadillac?" At that time, Cadillac meant a large, luxury image—all that. But the cars apparently didn't carry that image through. They were very good cars, great driving. The engineers loved them, but, unfortunately, most of our customers probably did not.

Author: If the fuel crunch had come in '86, GM would have won out?

HM: Yes. Some of the other companies did similar things, but at that time we found out that those cars were not what the public would view as a Cadillac.

Author: What was your involvement with the Northstar System?

HM: I was the vehicle chief engineer for Eldorado/Seville for 1992. I'll go through that and lead up to the Northstar involvement. It'll give you a better picture of what happened.

For the '92 programs, we truly got the voice of the customer, and I think we interpreted it correctly. We did focus groups with twelve to fourteen people at each session. They were composed of both customers and noncustomers.

We got together a group of current-model 1989–1990 owners who loved Eldorados to see what their opinion was, and a group of Seville owners. We even did a group of Fleetwood owners and got their opinion of it. We talked to Mercedes owners, BMW owners, and some of the Asian product owners to find out what we were missing.

We did this because the idea of the '92 Eldorado/Seville program was to start to penetrate into that luxury car market that was represented by the Europeans and the Asians.

As a result, we found out what people thought made a Cadillac. The question we asked was, "What makes a Cadillac a Cadillac?"

Some of us had some gut feel about that, but didn't have any proof of it. It turned out that people thought that Cadillac—and this is probably history—meant powerful engines, and that goes all the way back to the V-12s, and the LaSalle, and what have you. It also turned out that Cadillac meant V-8, when you got deeper into it; they didn't expect tens or twelves, and they certainly didn't expect sixes. V-8 and Cadillac were kind of synonymous terms to them.

Cadillac also meant technology, which was a little bit of a surprise to us, but they viewed the role of the automatic systems, the climate-control systems, the power

1983 Cadillac Fleetwood Eldorado Biarritz, production car. The grille contained narrow vertical bars separated into three rows. In addition to the brushed stainless steel front roof, the Biarritz featured wire wheels and Biarritz script and opera lamps on the sail panels.

seats and all that, as technology for Cadillac. Comfort and convenience were very high on their list of Cadillac hallmarks, as was safety.

Cadillac was considered a leader, but perhaps one of the biggest attributes was distinctiveness. Looking back, this was probably where we lost out in the '85 and '86 programs.

Surveys and clinics were done. With the designs, we had the challenge of the investment. We did clinics for the Eldorados and Sevilles. Something that would relate to the design of the vehicle, as well as the powertrain, is that when you do these clinics—and it takes about two hours to fill out all the forms—the bottom line is the question, "Would you consider purchasing this car when it comes out?"

There were five categories. The lowest was, "I would not consider purchasing it," and then there was, "I probably wouldn't consider purchasing it," the middle was "I'm not certain," and then there were two categories that were very positive and one that said, "I would probably consider buying this car" or "I would definitely consider buying this car." The rule in the clinical approach is that if 60 percent or greater say "I would probably definitely consider buying this car," you have a winner according to the marketing experts. The Seville had the highest marks ever. I think it was—I'm probably not accurate—over 80 percent that said that they would probably or definitely consider buying the car. That was a big plus. That was a winner.

The Eldorado wasn't as well accepted initially. We actually went through several fiberglass models before hitting a winning design, which is the design that is out there now. We got about that 60 percent figure for acceptance of the car.

One thing I should mention is that we never considered doing the new car and the Northstar the same year. Now, when I say never, there were some folks in Cadillac who did, but it was my recommendation that we not do that.

1983 Cadillac Fleetwood Eldorado Touring Coupe, production car. Note the flush hood ornament on the hood, trim changes, and special tires. The most collectible of all Eldorados of this body style, the Touring Coupe came with a beefed-up suspension, larger steel-belted tires, and a flush-mounted cloisonné hood medallion. Sonora Saddle Firemist and Sable Black were the only two colors used on this luxury touring machine.

We had had earlier programs where the car design and the engine design were brought out at the same time, and they had become dependent upon each other. The potential for delays in either of those programs affects the other. There's enough work to do when you do a new vehicle design that I don't believe you can handle the two programs going at once, and still do it effectively—getting the best out of both programs.

So, the Northstar was always planned to come out a year after the vehicle program. Some of the press indicated that the Northstar was delayed for whatever reasons. But that's because the press thought they would come out the same year. That was never a consideration, at least on my part. I am sure most of the Cadillac folks who were involved would agree.

We did something with the '92 program that we really hadn't done earlier. We formed vehicle teams. There were about seven or eight of them at Cadillac. Eldorado/Seville development was one vehicle team. This organization was one of the great reasons for its success—for both the car design and the powertrain. In developing a vehicle team you have representatives from every discipline.

Author: Are you talking about the mechanical part and not the design part?

HM: The whole thing. The whole thing was in one vehicle team and, again, we had representatives from styling.

Now when you think about styling and powertrains, you might not think they are synonymous, but there's a lot of work to do stylingwise on a new engine program. The appearance of the top of the engine, the color compatibility—you have to know if you are going to do all black, with just

1984 Cadillac Fleetwood Eldorado Touring Coupe, production car. This was the second year this de-chromed version of the Eldorado was offered by Cadillac. Also in the Eldorado line-up was a Biarritz convertible, produced under the auspices of Cadillac by American Sunroof Corporation (ASC), for $31,286—as compared with the year's Eldorado base price of $20,342. All Eldorado grilles for the year featured a vertical motif with the word Cadillac in the lower left. Body-colored moldings protected the sides of the standard and Biarritz models.

some accents, for example. But that's all taken into consideration. You wouldn't believe the amount of work that's done on the appearance of the underhood, and the amount of money that's spent to make it look good and be serviceable.

Author: You probably know of Frank Hershey. He was a designer in the early years—the Harley Earl years. He did the little Thunderbirds for Ford later. Once when we were with him, every time he'd see a Ford he'd curse and say, "Look at the way they just dumped this engine in the engine compartment." He said, "GM doesn't do that."

HM: There's a basic styling of the engine itself. In the old days the rocker arm covers, and now the cam covers, were considered from the standpoint of design—how they look in conjunction with other engine components, and even the routing of the hoses and the electrical harnesses. You just don't route them in there, you try to do kind of straight-line routing wherever possible. It's amazing how little things improve the looks of the engine; it makes it look orderly. And it looks like you spent a little time under there. The air cleaner covers are a key part of making the whole underhood arrangement first class. Plus, we consider the serviceability of it. You know, you want to get things where you can get access for service.

Author: So there's someone from styling actually involved in the placement of engine components?

HM: Right, we actually had a war room where everything was done and we met at least weekly and brought all decisions that had to be made. There was a roomful of twelve to fourteen people: styling, engineering (we always had two or three of us there), program management, financial, sales, material management. We had a manufacturing representative, public relations, marketing, customer satisfaction—basically every department at Cadillac and outside Cadillac that was working on the program, everyone involved.

It may surprise outsiders, but bringing all those disciplines together was not done in the past. It was not done anywhere to my knowledge. In the past, as you know, you would do the engineering part of the program and then pass it over the wall and then they would do their part of the program. But doing it simultaneously, as you can imagine, has tremendous advantages and it worked well for us. And we actually, as a group, did the roll-out plan for the Eldorado/Seville and the Northstar introduction.

We didn't used to do any early press rides until the vehicle was introduced or until within a couple of months of production of the vehicle, and we imposed some restrictions in terms of writing anything prior to introduction.

But the roll-out plan for that whole program involved the prints of the '92 auto show. We introduced the Seville there even though the '91 Seville was still in production. This was like seven or eight months prior to the actual introduction of the new car.

That was at about the same time Cadillac won the Malcolm Baldridge Award for quality. That was really part of the plan and, as a result, we got great press coverage for both the car and the engine. It worked really well for us.

Now I'll talk about the Northstar specifically. We truly listened to the voice of the customer. Again, it was pretty

1985 Cadillac Fleetwood Eldorado, production car. Pretty much unchanged from 1984, this was a "last convertible year," but there was no fanfare.

much power and V-8s—and responsiveness and the safety that's involved in that power. The packaging was a challenge for that engine because it went basically under the same underhood as the 4.9 liter. We spent a lot of time on that. Early on, it looked like it was a going to be a great challenge, but, again, we got all the disciplines involved—primarily service—to make sure the engine could be serviced. We were constantly together and eventually got to a design that could be serviced well and conveniently. This, in spite of the fact that when you look down through the underhood you probably can't see the ground very much, simply because we've occupied a lot of that space. It's a good, efficient design. We had a couple of versions of the engine and they are still out there: the high-torque, low-horsepower end of it, and the low-horsepower—I think it was the 275. The high-torque gave you a different launch view and that's the one that we put in some of the models. I can't remember which versions went into which models, but you know the Allanté was originally introduced with the high-horsepower engine, the 295. It had more of a European, linear, off-the-line feel. You know what I mean? You've driven any Mercedes, for instance? You get the feeling that even though they get 8.5 seconds to 60mph, you don't get the full sensation because you don't get a quick launch. So anyhow, we think it was tailored to the vehicle.

The roll-out plan was to first put the engine in the Allanté, which we did. This was the technology leader. Later, we wanted to put it on the STS Seville and the touring coupe version of the Eldorado the following year. The third year of the Northstar, it would go into all Eldorados. And, the third year being 1994, all Eldorados, all Sevilles, and then the Concours would have it. That's where we are at this point and I don't know what their plans are now because I retired almost two years ago.

But, I can remember the first drive of the Northstar System in a prototype Eldorado. The prototypes are typically kind of little cars, you know they don't have all the comforts in the early renditions. Even at that point you could feel the great aspects of that engine, and a terrific 0 to 60 which probably was eight seconds at that time. The 60 to 80, the capability of passing, and the performance when you stepped into it at that speed was something we had never felt before. Then, for me, at least, it was new to drive 150mph and have the stability of that engine and of the overall vehicle—and the lack of lift, etc.—in that Eldorado. It was a thrill, no question.

Author: Ruzzin was telling me about driving the car that fast in Germany. Do these cars come with tires that are okay for that speed?

HM: The STS and the Eldorado Touring Coupe are cars we have fitted with tires that can go 150. The base versions do not have them; that's always the restriction, basically, the tires. Even in past years the engines would go 127 to 130mph, but we did not put tires on the cars that would handle that, so they would hold back the engine.

Author: What were the basic differences between this Northstar engine and the previous one?

HM: To me it was 0 to 60 and the feel of that power. It was the 60 to 80 performance. You know, usually you can design in a pretty good launch field 0 to 60, but you might run out of power by the time you do 60 to 80, or 60 to 100 or 110—just to move you out and allow you to pass. There's safety in allowing the driver to get around somebody in a situation where the area that you can pass is somewhat restricted.

Then again, none of us, at least the fellows that I was working with, had driven cars at speeds of 150-plus. It's amazing, you start to get the feel for how the Germans, for instance, can get used to that. Once you have confidence in the vehicle, you adapt surprisingly quickly to driving it at those speeds over 100mph. The car was still quiet and it was still fuel efficient.

One of our advertisements was that they were able to drive the car something like fifty miles with no coolant. That's just an indication of the great efficiency of the engine.

The powertrain guys could tell you more about the volumetric efficiency, the dual overhead cams, and the relative quietness of the engine despite its power. It was a total system; for instance, the transmission had to be new to carry that kind of torque. Northstar is truly a great overall system.

AMERICAN ♦ CLASSICS

Chapter 6

1986–Present: Mistake and Masterpiece

It's difficult to believe that two decades have passed since the Arab Oil Embargo. In 1974, Americans who hadn't suffered real shortages in anything since the rationing years of World War II suddenly found themselves sitting in endless lines at gas stations across the country waiting for a few gallons of fuel. Sales of big cars were immediately affected. For example, Eldorado sales went down during, and for a short time after, that period. People started looking closely at gas mileage figures, vehicle weight, and politicians who would promise oil self-sufficiency for the U.S. Gasoline went from around 50¢ per gallon to something approaching $1.50.

After the embargo ended, it slipped farther and farther to the back of people's minds. While the embargo did create a greater market for small foreign cars and compact American models, most people were still surprised when Cadillac was brave enough to moderately downsize its full-size cars in 1977. It was a jolt to American sensibilities to see Cadillac lose so much weight, but the division managed a record sales year with this fresh approach.

Imagine the concern at General Motors when their energy experts, in 1979, began to forecast gasoline prices of $3.00 or more per gallon by the mid-1980s. Most likely, these predictions were set off by the Iranian oil shutoff at the end of the 1970s. GM executives thought something drastic had to be done. Suddenly, Cadillac's downsized 1977 models didn't look so small anymore. Besides, Japanese and European imports were now a serious competition problem, and too many Americans were looking to these foreign suppliers for greater fuel efficiency and reliability.

The Cadillac Motor Division, even though its name was synonymous with opulence and luxury, was going to have to chop their cars down and increase gas mileage. At the same time, Cadillac was going to have to *somehow* maintain its identity as a luxury automobile.

To put it briefly, this was the scenario that led to the introduction of the downsized Cadillac Eldorados of 1986. The decision ultimately placed the marque in danger of extinction. It was a mistake, but one made clear only by the 20/20 vision of hindsight.

The birth of the 1986 Eldorado was a difficult one.

The general manager of Cadillac wanted to downsize drastically to allay concerns for the oil crunch he believed to be on the horizon. Everyone associated with design staff, including Irv Rybicki, Chuck Jordan, and Wayne Kady, fought these changes. Their main concern was that the Cadillac would loose its character. They lost the fight, however, and Eldorado went through a period of eclipse from 1986 until the introduction of the 1992 model.

This story is told in detail here by several designers and engineers who lived it. The model changes are described in the accompanying photographs.

Great things can come out of dark days and setbacks. Under John Grettenberger, Cadillac found its way out of this darkness and into the sunlight of success and critical acclaim. One wonders if the Northstar system and the world-class 1992 design would have been as great without Cadillac Division having gone through such difficult times.

The 1988 Facelift of the 1986 Eldorado

Chuck Jordan having returned from a Southern California press conference for the 1986 Eldorado, came into the studio asking what we were doing to fix the Eldorado.

We all knew this had to happen, not only was the car made smaller, the total character had changed. All the design cues that were part of Eldorado's heritage were gone and the down-sized car looked like a used bar of soap.

We directed Chuck to the back of the studio, where we kept the master clay model. We had already made changes to the rear quarter, taillamps, and rear bumper, adding length to the quarter with a symmetrical taillamp—reminiscent of the previous Eldorados of 1979—'85. We felt that the rear of the weakest—besides that was all that the budget would allow.

It was felt by everyone that this was not enough, and there was an urgent move to get more funding to change the front, to add more Eldorado presence, and balance for the changes made in the rear.

Timing was also a problem. But once we figured what we were able to do, there was great cooperation with the engineering and die people to expedite the changes. A new hood outer panel, fenders, and grille were added to complete the changes. All that effort helped some, but it was only a Band-Aid fix, not nearly enough, and too late.

Wayne Kady
September 26, 1994

Julian W. Carter: Notes on the Philosophy Behind the 1992 Cadillac Eldorado Interior

Julian Carter.

Julian Carter received his degree from the Cleveland Institute of Art in 1973 and joined GM Design Center in 1974 as an Associate Designer in Cadillac Interior Studio. He has had several assignments at GM including Chevrolet, Pontiac, and the '85 Corvette Indy Show Car. Carter rejoined Cadillac Interior Studio in 1988 as Assistant Chief Designer for Marvin Fisher. He is now with Pontiac Studio.

This interior had to hint of its heritage as a flamboyant domestic luxury coupe. I feel personally that it was the flowing "devil may care" line on the door and rear compartment—the way that it wrapped around from the upper instrument panel, swooped to just above the armrest, then rose to continue onto the package shelf over the rear seats—that forged a link with the Eldorado's past.

In addition to all the wonderful functional attributes of the seats, they had to be a very distinctive design. This started with the specialized headrest shape. This was especially so in the rear compartment where they blended sculpturally into the rear speaker housing.

"Substance" was to be an ingredient as well. Gone was the look of minimalism and flatness which characterized the previous generation of Eldorado seating. Form was added or deleted to enhance ergonomics and comfort.

The seats were to possess some uniquely American design cues, such as the controlled wrinkles.

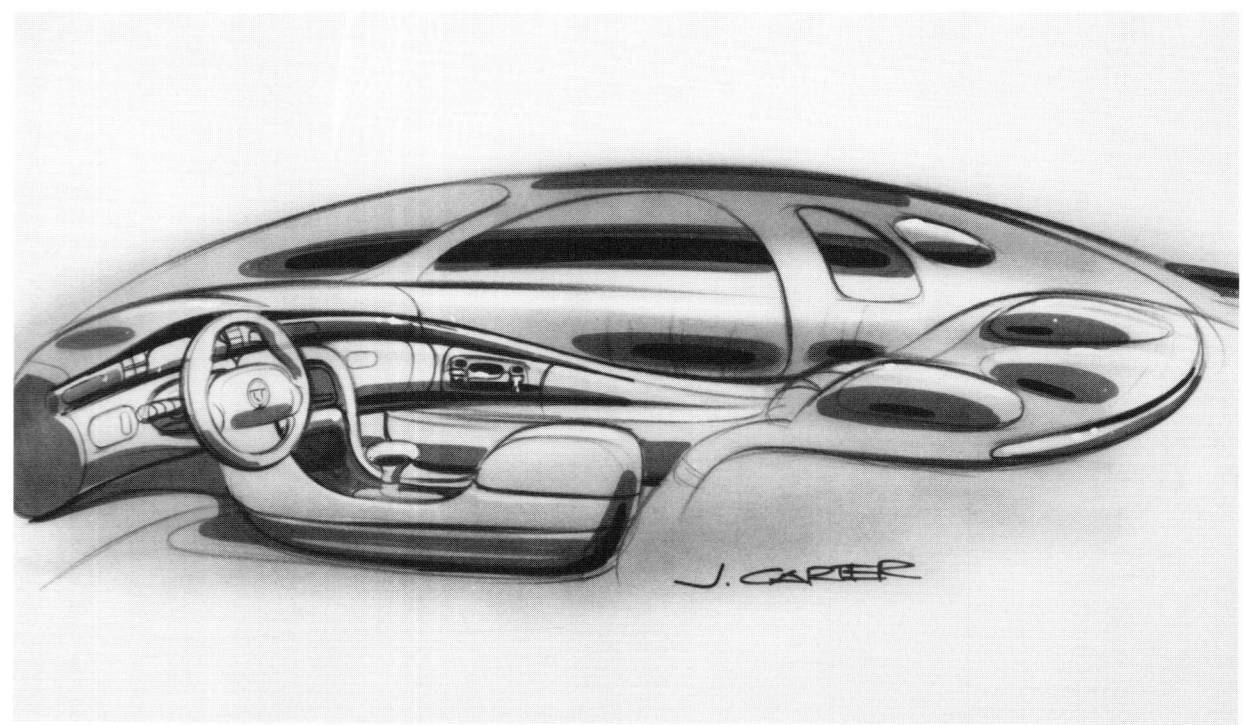

Early Julian Carter rendering of 1992 Cadillac Eldorado interior.

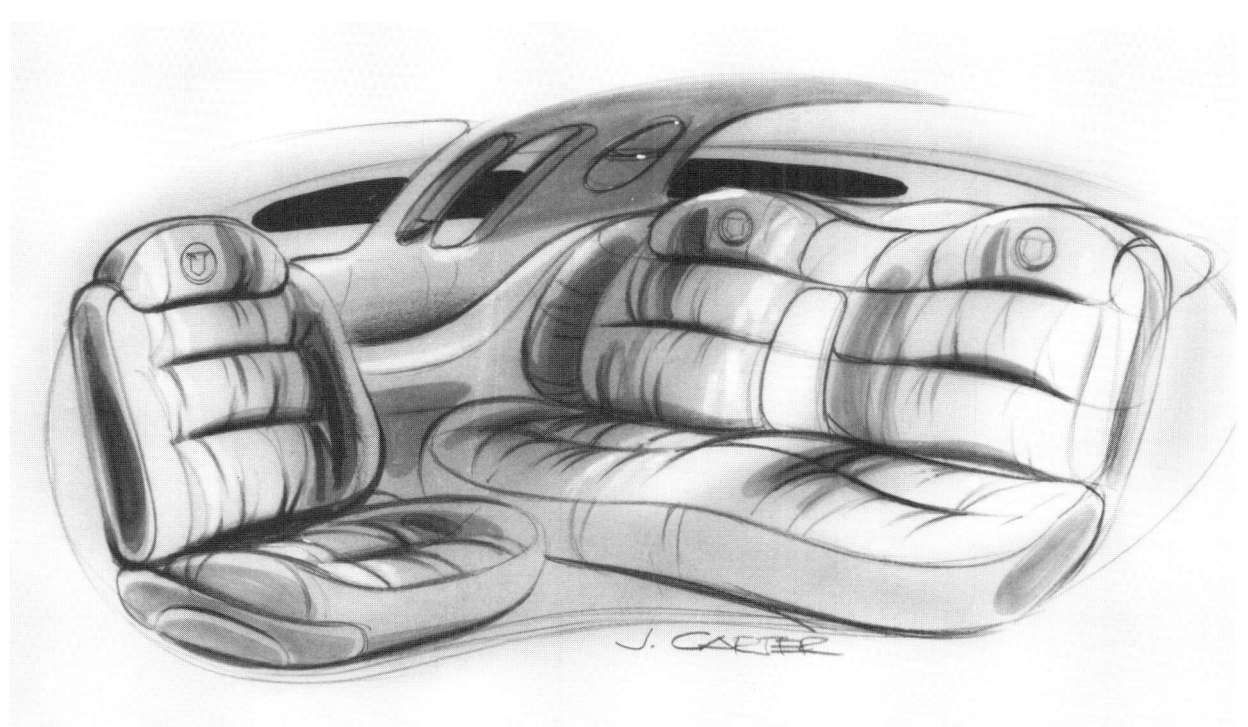

Carter sketches for a front seat and the rear seating of 1992 Eldorado.

1993 Cadillac Eldorado interior—'92's was very similar.

DICK RUZZIN INTERVIEW
The Design Story of the Current Eldorado and Seville

Dick Ruzzin joined GM the summer of 1959, but started with Fisher Body (for a couple of years) before joining Design Staff by presenting a portfolio and being hired as a designer. He first worked with Cadillac in 1986 as the chief designer at Cadillac Studio, responsible for the current line of Cadillacs, including the Eldorado and Touring Coupe, Seville and STS, Fleetwood, DeVille, and Concours. Ruzzin is currently in Germany at GM's Opel plant where he oversees all GM international design.

Dick Ruzzin.

Author: Is the current Cadillac the first one you worked on?
DR: Yes, when I went to Cadillac, we started the Eldorado and the Seville, the '92s. And we followed those with the Fleetwood, the big car with rear drive. We did the Concours and the DeVille which were most recently launched. So all of those cars were done at Cadillac Studio under my direction.

Author: What is your concept of what the Eldorado should be?
DR: If you go back and look at the previous Eldorado and Seville, you'll see that the pendulum swing to downsize had swung to the very bottom and the pressure on the corporation to meet the CAFE (Corporate Average Fuel Economy) requirements was very strong. To put it in a simplistic way, the government in Washington had gone to General Motors and said, "Look, this is what we have to do. We need someone who is a good social citizen and someone who will support this. If General Motors will support these guidelines, then the rest of the industry will have to also." So, the powers that be at GM at that time naturally wanted to be good social citizens and they said, okay. Then the fuel economy rules went into effect. To make that work it meant that the cars at the heavy end of the line—of which there were a great many at that time at General Motors—had to have dramatically improved fuel economy. Earlier, fuel economy technologies had not been developed; nobody had really cared about it—by that, I mean we had found it wasn't too important to the customer. One of the great engineers at Cadillac, Hank Meyer, told me that in those days when the downsized Eldorado was done, there was actually a discussion about a three-cylinder Eldorado!

That kind of shows how far things had gone off center: that, in a meeting when they were looking at engine choices, they would even discuss a three-cylinder engine as a possibility.

Author: So there was a lot of discussion about how small the engine in the downsized Eldorado should be?
DR: Yes, originally they were going to have a six-cylinder engine in the car. Then at the very last minute they decided to put a V-8 in it. This meant that since the car designs had already been released for production, in the computer they had to cut them open down the middle and spread them out so they could put a V-8 in. You know, you're talking about a transverse engine. If they hadn't done that, there's a good chance that Cadillac as we know it today would be gone, because they would not have sold anything! It would have been absolutely dead.

That's the background on the downsized Eldorado. But I thought the mistake on that car designwise was that the design didn't emulate the previous Eldorado. It emulated the C-car instead. If you look at the downsized Eldorado, you'll see it has very few Eldorado design cues, especially in the body's side. It looks like the Cadillac C-body coupe that came before it.

When I first got to Cadillac Studio, they were just completing a facelift on the downsized Eldorado where they were changing the rear quarter, the bumper, the grille, and a few other things. This made it look much more Eldorado-like, especially when they added the vertical taillights. Actually, that facelifted downsized Eldorado was a pretty neat car. It was compact. It was pretty darned nice, especially when they increased the engine size and improved the suspension.

Author: Did they change the length, making it longer?
DR: No, I don't think they changed the length, but it clearly looked longer. That was a question that was often asked. It confused people. Pulling the taillight out, making it more of a fin in the back made the car look longer from the side. Then if you looked at the bumper in front in plane view, it was very well executed—and I didn't have anything to do with it—but when you looked at it very closely it was very strange, the bumper didn't follow the sheet metal. But this had the effect of pulling the front corner out, making the car look more bulky. Since the body looked more bulky, this made the roof look smaller. All this was done in a classy way and it was much more Cadillac in character. The car had a much better Eldorado feel than what it had originally been. I remember having a couple of black ones and I thought they were very nice cars, with their leather interiors and everything.

Author: So you began work on the new Eldorado?

DR: Yes, after the facelifts on the downsized Eldorado we went through about five full-size car designs looking for what would be next. We were trying to find our way. We did all these cars and had clinics and there was a winner—a marginal winner. That was the Pininfarina model, the only one with a wide body protection molding. The lack of this protection was a deficiency in all of our early models.

Author: Let's clarify what you mean by "a clinic."

DR: Clinics are an ongoing, developing art and science. A clinic is gathering up a number of people—three or four hundred—from one or more parts of the country and having about two hours' worth of questions. Then having them sit down in small groups to have carefully orchestrated discussions about certain subjects, trying to get some information firsthand from people. The really important information comes from the forms they fill out, because then you can average things out. You can see what really works and what doesn't, statistically.

Larry Erickson, GM Designer who worked with Ruzzin on 1992 Eldorado

Author: In these clinics, are you showing a picture of the car, parts of the car, or what?

DR: We're showing a full-size fiberglass model or photos. Many different kinds of properties can be used.

I'll get back to the story. Finally, they did a clinic where they wanted to critique effective color on a design. They took the Eldorado, the Riviera, and the Toronado at that time—this is for the future model. They painted them dark colors: one was dark maroon, one was dark blue, and one was charcoal, I think. These were all the new E-cars, the new coupes. And they all bombed. All three of them bombed.

When that happened, I remember Chuck Jordan came to me and said, "We have a chance to do one more Eldorado." He almost said it in a whisper, like we're going to have to go through this again. You know, because we had just done so much work. Keep in mind that we were at the same time doing the Seville, an equally difficult task.

Author: So you're talking about another Eldorado proposal?

DR: Yes. I'm talking about the Eldorado that ended up being the one on the road now. So we had one more chance to do another one. We had two fiberglass models. Jordan had a good idea. He said, "Go work someplace where there's nobody around."

At that time we were just trying to find a design, and we had developed a lot of tricks using paper and cardboard and tape. We got so that in an afternoon we could make a car (a different design) out of paper—on a real car. You know what I mean, just by moving our materials around. We got very good at it. So it ended up that we took those two fiberglass models into another part of the building. It was Thanksgiving vacation time and I was off for a couple of days and came back to find the cars already started. Larry Ericson, who later became my assistant, kept saying that a line on the side of the car was wrong—Ericson is the designer of ZZ Top's car, *Cadzilla*, which he designed right in Cadillac Studio. We'll never get an Eldorado unless we get rid of that line. The line he was talking about was a line that was on practically every model that we had done.

What Larry was saying was that every model we had done was wrong because it had that line on it. So the first thing we did when we started using paper and cardboard was that we threw that line out and did something different. After a couple of days we had created this model out of paper and we showed it to John Grettenberger, the head of Cadillac. (Of course, Chuck Jordan and Dave Holls saw these proposals first.) It was decided then that we should do a clay model. In just eleven days we went from something in paper to a full-size clay model! You don't know the significance of that, but let me tell you, in the best of circumstances, if you do that in a month, or even six weeks, you're doing great!

So we had the model all done and we had just one chance to take it outside and look at it before we cast it in fiberglass. The next day was to be a Cadillac meeting at Design Staff and we were going to show it to them. We had looked at several Eldorado proposals on the patio, but here it was: it was fall, it was cold. We were out there wait-

1986 Cadillac Eldorado full-size clay model photographed on the turntable on the patio of the GM Tech Center in March 1981. This model was Cadillac's response to the prediction that gasoline prices would reach three dollars per gallon by 1986. Wayne Kady said that one of the problems in being ordered to produce such a design was that the car tended to get very rounded as it was made smaller and it was difficult to maintain the Cadillac Eldorado's distinctive design cues.

1986 Cadillac Eldorado, scale model in clay. Although Cadillac management insisted on downsizing the Eldorado and other cars in the 1986 line-up, there was strong opposition, especially in Design Staff. Irv Rybicki and Wayne Kady, among others, kept pressing for some sort of compromise solution that would maintain Cadillac's character and prestige. After the previous model was presented, this clay scale model was shown to management as a possible compromise for the Eldorado. Many of Eldorado's design cues are here: the elongated hood, the horizontal grille, and an innovative overall design that could have easily assumed the Eldorado mantle. In this rear three-quarter view, one can see the solid design concept. It doesn't seem to matter that the sail panel is quite narrow, because it complements the rest of the concept. The mid-body beltline tightens the design, giving the car that feeling of elegant sportiness that Chuck Jordan said Bill Mitchell wanted in the Cadillac. This concept might have saved Cadillac from the Dark Ages of the 1986 downsizing had it been adopted.

1986 Cadillac Eldorado, Fiberglass model. Styling increased the size of the sail panel seen in the last full-size clay model, but, unfortunately, this was pretty much the Eldorado design that went into production for the 1986 model year. Had gasoline prices leaped to the three dollars per gallon mark predicted, Cadillac might have been declared the winner in the luxury car market for the time, as Lincoln was still selling its large, dated Town Cars and its recently introduced Mark VII. But the price increase did not happen and Cadillac had failed to maintain its leadership in design. It was difficult for the designers to put that Eldorado name on the front fender, but they had no choice.

This is the 1986 Eldorado that went to market. Cadillac offered it with available touring suspension and wide body-side molding. When the public saw it on the road, they had to look twice to believe it was a Cadillac.

ing for them to bring the car out. We had done the car in a silvery color and it was a very dark, gray day. It was about 4:30 in the afternoon and starting to get dark. Finally, the car starts coming out. We had brought out so many other proposals before, only to get them out and have them not meet our expectations. But when this car came out, it just looked smashing! It looked fantastic and we thought that this was really going to be it.

Just then the sky split and a sunbeam came down, right on the car! It was really weird! One of the guys next to me said, "Oh, it's a sign from heaven!" And he crossed his fingers in front of him, like when you see an apparition in the movies to ward off the devil or vampires, or whatever.

The car really looked great. We took it back in and didn't change anything for the presentation to Cadillac the next morning at eight o'clock.

That night, I went Christmas shopping—for the Christmas of 1988—and I got to thinking, you know, it had always made me kinda upset that the Cadillac people would always come so early in the morning. They'd wander around and see everything, and by the time we made our presentation to them it was kind of a letdown.

I thought: what should we do? I decided we should put a cover over the car, so I bought some Christmas wrapping. We put a green cover over the car and made a big bow out of Christmas wrapping.

At that time the Cadillac people didn't know me very well. I'd only been there for about four or five months. I didn't know them very well. I didn't know how they'd take to a joke, but here this thing sat. They came in. John Grettenberger said, "What's that?" He knew that we had done another Eldorado and he knew that we were running a month late for the car's release.

I said, "Well, this is your Christmas present. Do you want to open it now or later?"

He said, "No, that's OK. Leave it till later."

So we went about our other business. You have to understand, we had gone through so many design proposals by that time that the people at Cadillac and the designers had developed pretty strong ideas as to what the car should be. Every time you do another proposal, they have another opportunity to test their theory against what it is you're showing. So everyone has a constantly ongoing idea about what the car should be. This is very bad, because as time goes on, you reach a situation where there is no answer.

We got through with our business and pulled the cover off. There was just total silence. Everyone was stunned. All I heard was feet shuffling. Everybody was impressed.

Author: So, that's when you did the fiberglass model and took it to a clinic?
DR: Yes, we did the fiberglass model, put it in a clinic, and it was a huge success.

This was a difficult thing because we were doing two cars at once. We were working on the Seville at the same time. That is very difficult for a small group like ours, even though we had an awful lot of extra help from the Advanced Design activity.

Author: You finally had a winning design?
DR: Yes, but it was a really tough thing because the Eldorado before the downsized one was so well received and so well liked by their owners that they thought there was no way you could make a better one than that. They thought it was absolutely the perfect Eldorado. We had to run against that, which was difficult.

Author: Didn't the exterior and interior designers work together more closely on this project than on others?
DR: Yes, but it was no big overt thing. The interior studio was together with us in the same room with a partial wall between us, and we all worked and talked together everyday.

Did you know John Selznack, who did the instrument panel, is over here in Germany for a period of time?

Author: That's the first time I've heard that name.
DR: Yes, he was the designer who did it. He established the theme, the wraparound concept with the wood inlay. Now, Marv Fisher was the chief interior designer. He's a fantastic designer, but John Selznack is the kind of person they'll put in a studio when they want to start something. We already had an instrument panel for the car left by the previous chief interior designer, Drew Hare. Selznack was working on an instrument panel for one of the other Cadillacs and had it all laid out on the board full size so that it extended into the doors. All one long thing. It was so neat that I said to Marv Fisher, "Why don't we use that in the Seville?" Marv doesn't talk a lot, but he said, "Yes, that might be good." About a month later, I noticed he was using it for the Seville.

Sometime later, we were trying out the new instrument panel. They had put it in the old downsized car and we were driving it around. I took it home one evening. The old car had this very boxy instrument panel with a horizontal line at the top. What happened was that this horizontal line was a reference to the horizon. As the car jounced, that horizontal line amplified movements of the car. What happens when you go to that curved shape in front of you that the Eldorado and Seville now have, is that it rolls with the horizon. So it's not a reference to the horizon anymore because it's a curved surface. It takes away some of the jouncing feeling when the car moves down the road. It tends to make the ride feel smoother, the car more stable.

So nothing of the old design was used. Marv Fisher came in there and brought all the parts together—such as what Selznack had done—and made it all work. He's a super designer.

In the end, in the Seville and Eldorado, they have much the same interior, you have a very harmonic, quiet design that is easy to live with.

Author: As a designer who worked on both, what do you think of the current Eldorado and Seville?
DR: That's an interesting question. For some reason, I have absolutely no favorite between the two. They both do a specific job. The Eldorado appeals to a more traditional American coupe buyer, not someone who would consider an import. These people want something that doesn't go with the mainstream, something unique. They don't want a car that is round like all the Japanese cars, or whoever else is making round cars. We ended up giving them that kind of a car, the car they wanted.

Coupes are kind of going downhill. Our coupe market is quite flooded, but the Eldorado seems to hit a certain level and seems to be doing very well. I know the Lincoln Mark VIII is also doing well, but I've heard they're practically giving them away.

The Seville is a different sort of thing. It is a real American version of an international car. If that's the kind of car you like, then it's perfect for you.

But the two cars do two totally different kinds of jobs. I guess that's why, for me, it's not one or the other. They're both a certain kind of answer for a certain kind of problem.

Author: I'll ask a devil's advocate question. Some people have said that this current Eldorado looks somewhat like the '85 Thunderbird from the rear three-quarter view. What do you think?
DR: First, I'd say put the two cars together. When you do, you'll see they look very, very different. Both are coupes. Both have a vertical rear side window shape—something you just can't walk away from on an Eldorado. This is one design element the car carried forward through the years. Now you could take the current Eldorado and change that, and you'd be OK. But when we came out with that car, it was too soon to walk away from that. You just could not do it.

The Seville is extremely harmonic, that's something they say over here. It's like notes in a musical piece. There's nothing there that doesn't belong. Everything works together.

There's something unique about that car. I remember when we introduced the car in LA and then took it to Phoenix to do pictures, that was the first time I had seen that car driving. It was very weird, because I suddenly realized that the appearance of the car as it was driving down the road was affected by the electronic suspension. The car has that strong line through the body. That isn't present on the Eldorado—we were trying to make the Eldorado look different, of course. Look at that line when you see a Seville going down

The Biarritz name continued to be just a trim package for 1987. The car received larger standard tires, some changes in the suspension system, and engine damping, but, overall, few changes were made for the 1987 model year.

the street; it doesn't move, or tip up and down on the ends. Normally, cars going down the street will jounce front to rear. But not the Seville; it has very stately, elegant motion because the body is kept very quiet in its movement by the electronic suspension. When I saw that, I knew we had a real winner.

It's an interesting thing. People who want to see a European car in it see it. People who want to see an American car in it see it. It somehow has a blend. You'll read articles both ways. I see them over here in Germany and the rest of Europe quite a bit. Actually, they look even better over here because they're more of a standout. There is just not the great variation in the size of cars over here as in the States. In the States you have the small ones like here, but then you've got much larger ones. Whereas here, it stands out in traffic in a more dramatic way. The engineers at Opel have told me more than once that it is the only American car they have ever really liked.

Author: Have you driven a new Eldorado over there?
DR: Yes. And I've seen a lot of Eldorados over here, too. The difference is that it is not a European car at all. It is American. That is what we were trying to do. That's what the research said the Eldorado had to be.

What's amazing to me is that the engineers were able to configure the ride and handling so that the cars work so

1988 Cadillac Eldorado, production car. Without the expensive changeover to a new body style, Cadillac tried to respond to complaints about the styling of its downsized cars. It is difficult to recognize the new power dome hood, front fenders, sail panels, and, even the new rear quarters and deck in a photograph, but you can see the improvement when you compare a 1987 with a 1988 in the flesh. Now the line from the rear deck is carried forward to the back of the rear side window giving the car a more solid look. This marked a return to a traditional Eldorado design cue. Otherwise, the visual effect and real changes for the 1988 model year were minimal. The 1989 has little changes from the 1988.

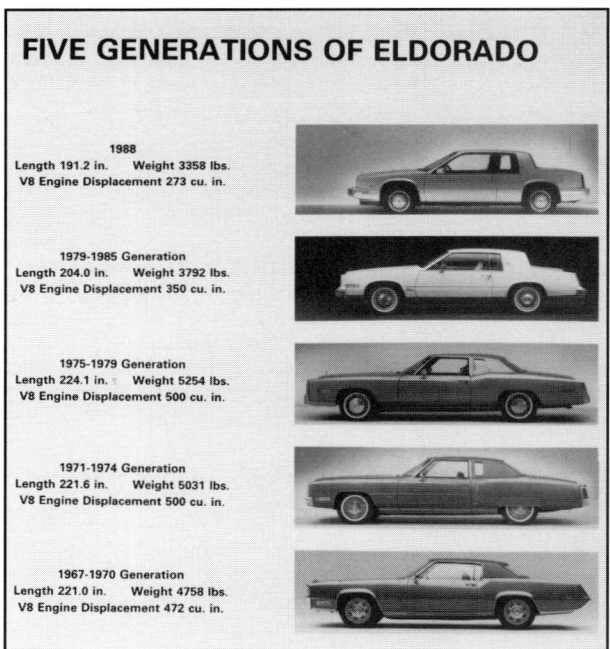

Because this was the 20th Anniversary of the front-wheel-drive Eldorado, Cadillac thought it necessary to release this comparison of Eldorado front-wheel-drive cars. Every car in this comparison is instantly recognizable as a Cadillac, except the 1988 at the top.

well on European roads. In driving both here, there is very little I would consider changing in the ride and performance. Here the car goes 155mph, handles great, has good aerodynamics, has good fuel economy, and everybody looks at it, but in a positive way.

In a Seville once, I noticed a cab following me in Frankfort. The men inside were motioning me to pull over. I didn't know what I was in for. It turned out they just wanted to look at the interior and the Northstar engine. You have to understand how far we've traveled to have that happen: to have a bunch of Germans want to pull you over to look at an American car's engine. When we started out on these cars, we would never have dreamed we could reach this point. American cars had just gotten so far down the scale of desirability, especially from a technological standpoint. To the Europeans, before the Seville, there was just nothing that was admirable about American cars.

Author: Now, the question everyone who reads this is going to want to know is, why in the world did GM send you to Germany after you were so instrumental in creating two of the most beautiful Cadillacs ever designed?
DR: Well, actually, we went through the whole Cadillac line before I left. We did all four cars over. What happens after that is some facelift work or whatever. Then they start over again. There would have been a couple of years I wouldn't have had much to do except work on facelifts. You see, if you are one of the few designers who can conceive cars, you are at the top of the list of people capable of putting a car together. For others, it might be doing facelifts. So I've been here for a couple of years and we've been going through the same thing here for GM Europe. We've done quite a number of cars already.

Things are different now at GM. When Harley Earl, and later Bill Mitchell, was head of GM Design, he was in charge of GM Design around the world. But when Irv Rybicki took over, they took the international responsibility away. I have complete responsibility for all design outside North America. That includes Australia, Germany, Saab—everything. It's sort of unwritten now, but this is all part of GM reorienting itself to the world.

The one thing I regret about coming over here so soon after doing those cars is that I missed the American public's reaction to them. I missed all that.

Author: They have certainly been well received. Do you have any sketches or pictures from your work on the current Eldorado?
DR: No. As chief designer you don't do many sketches. You rely on your people to do that. If you start coming in with sketches, you're likely to make people angry. It's best to make suggestions. Sometimes I would make doodles and things just to help my thinking process, but that was about it.

Author: What do you think is the future of Eldorado?
DR: I really don't know. Coupes are funny things. Right now, there's a lot of coupes in the market in America. The market gets crowded. Then someone says let's not do coupes anymore because the market's crowded and we won't get our money back. Then all of a sudden there aren't any there anymore. Somebody says let's do a coupe because there aren't any there. Then that's a big success and everybody gets back to making a coupe. Right now there are a lot of them, so coupes could diminish.

But that would really be hard, because now the Cadillac Eldorado and the Lincoln Mark are American traditions. I don't know what is going to happen.

Author: Do you consider the Eldorado the flagship for Cadillac? Is that the way you felt about it when you were doing it?
DR: It was when we were doing it, but I think what has happened is the flagship now—by consensus—would have to be the Seville STS. The coupes were always a problem. They were sporty, but they were two-doors. As people became more enamored with these performance sedans, they decided these cars had everything. Before they would have had an Eldorado and Seville on the market with the Eldorado outselling the Seville, but now the Seville STS by itself may outsell the Eldorado.

The sport sedans have taken over from coupes. It's very logical. You can go out with your friends and you don't have to worry about getting in and out of the back seat—you can still have a good-looking, sporty car. There was a day when a sedan could never look as good as a coupe because somehow the coupe was alluding to a two-passenger car. Everybody wanted to have a sporty image, and couldn't.

When we were doing Eldorado and the other current Cadillacs, we had a great many things against us. Everyone knew the Japanese were coming out with the Lexus and the Infinity. There was going to be a direct measurement against the capability of the designers at GM to compete against these formidable world talents. That was one side.

The other thing was that you were dealing with an American icon. You weren't doing a car like a Chevy or a Pontiac or a Buick. This was something that was part of the lives of Marilyn Monroe, Elvis Presley—I mean, the reputation of Cadillac goes way back. We were going to this expression, maybe the last expression, if it really didn't go, of an American automotive icon.

Author: You were sticking your neck out?
DR: That's right. That's exactly it. I didn't volunteer, but later I could see that that was the position we were all in.

But the third thing was that we were not only American designers competing with designers across the world, but there was also this feeling that GM design had somehow fallen by the wayside. So we had to prove that we could do a car as good or better than anybody else. I think we did that.

All this was going on in the back of everybody's mind who was working on the car. And there weren't that many, there were just a handful of people, creative people. And here we are doing this thing in an incredible rush. Trying to come up with something super special.

But Cadillac Division and John Grettenberger really

For 1990, Cadillac tried to repeat its introduction of a Touring Coupe model as it had in 1983. This model featured a 4.5-liter, 180bhp transverse-mounted V-8 engine, a retuned touring suspension package, 16x7in forged-aluminum wheels fitted with Goodyear Eagle GT+4, P215/60R16 tires, four-wheel anti-lock disc brakes, and a four-speed automatic transaxle. The interior had full leather seats with six-way driver/passenger seat adjuster, power recliner, and lumbar support; monotone rather than chrome accents, and more wood trim. Limited optional equipment consisted of a choice of either Delco/Bose cassette radio or compact disc player; astro roof with express open feature; and a theft deterrent system with automatic door locks. On the exterior the Touring Coupe was distinguished by reduced chrome moldings, grille mounted wreath, dual muffler outlets with bright chrome tailpipes, and crest emblem, body color door handles and exterior side mirrors, and modified export taillamps. Rocker panel and rear side reflex moldings are also revised from the base Eldorado are exclusive to the Eldorado Touring Coupe.

The 1986-style downsized Eldorado bowed out at the end of 1990. During its production time, all manner of cosmetic modifications had been tried on the design, including continental kits and special striping. These cars were comfortable and drove well, but many critics considered them incapable of carrying on the Eldorado name.

handled the situation very well. Grettenberger knew what he wanted it to be: not what it was, not what it would look like, but when the design team was finished, he knew he would know when it was right. He had very good direction. For example, he never allowed us to do those wheels that look like they're twisted. We tried to do some of those, but he never let us do it. He said he wanted classic wheels. While everyone else was doing the twisted wheels, we did the classic ones. We did a lot of beautiful wheels. Cadillac didn't bow to the trend.

You have to realize that Cadillac Division had lost its confidence in Design Staff. As I understood it, there had been an adversarial relationship between GM Design and Cadillac before I took over the studio. I made it part of my mission as chief designer of Cadillac Studio to break that down. I worked very hard at this and, I must say, very successfully. When I left, the relationship was the best of any division within GM. I am as proud of that as I am of the cars.

More importantly, John Grettenberger came in and literally restructured the whole division, and design was one of those elements. He really has a wonderful management team.

And, he has a good eye. He once looked at the clay model of the Seville, just in passing, and said, "It's a classic before it is even introduced."

Grettenberger is quite a visionary and knew, at the end, where he wanted to be. After about a year, I could see that everything he wanted was a very easy job for me to do. I knew what would please him.

Author: So he wanted something that looked like a Cadillac?
DR: Yes, but the problem was that no one had ever done one that had to do what this Cadillac had to do at that particular time. One problem was that Cadillacs had always been tradition bound. When you didn't have to worry so much about the size and weight, it was easier. But when you gave that away, things got difficult.

You reach a point where you need real substance. We provided the design and Bob Dorn the technologies.

The head of advanced engineering came to see me the other day and asked me if I had looked at the latest issue of Automotive Sports, a big German magazine. He said the aluminum Audi car was featured in it, and that in the same magazine was an Eldorado road test. He said that the Eldorado weighed ten kilograms less than the Audi with the all-aluminum V-8! He said, "Nobody says anything about this! This is fantastic!" It is.

The Seville is also a very light car for its size.

Author: There is still some prejudice in the automotive press against American cars.
DR: Yes. In fact, there was recently a comparison road test published in a German magazine which included the Seville, the Lexus, the Mercedes, and the BMW. At the end of the article it said, "To those of you reading this article, who work in the German automotive industry, take note. If this car is indicative of what is going on in America, things are no longer what they used to be." I thought that was very well put.

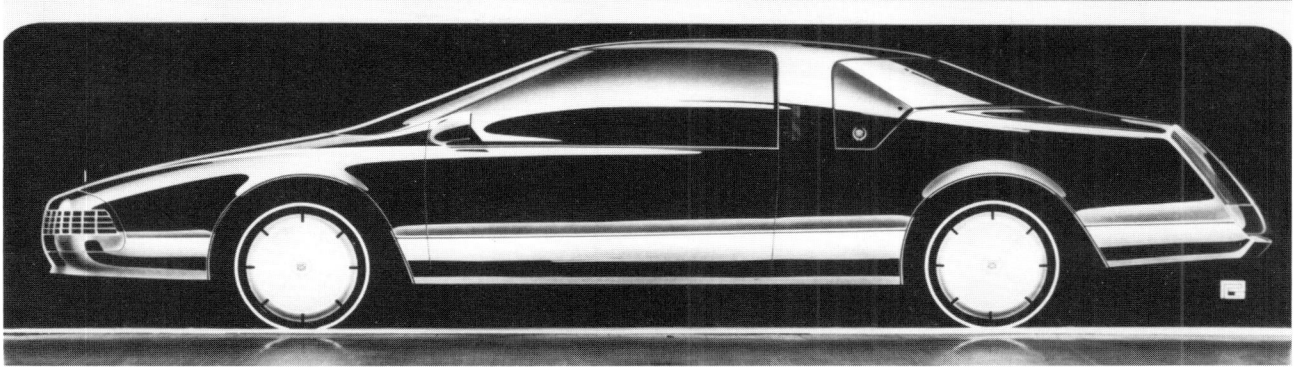

The three photos above, design models for the 1992 Eldorado.

Author: One last question. What was your favorite Eldorado before the current one?
DR: Well, they were all such big cars and I'm a little-car person. I mostly worked on little cars. But, if I had to choose, it would have to be the '67 Eldorado. The first one with front-wheel drive. It was a little smaller than the others. I had worked on the Toronado, Oldsmobile's first front-wheel drive that came out the year before. When we had our review of the design, all three of those cars—the Eldorado, the Toronado, and the Riviera—were reviewed at once.

What we saw then when we looked at that Eldorado were the standard accepted Cadillac cues of the time set on this dramatic profile. It was a very interesting car. Wayne Kady has a black one with a black interior that is quite beautiful. It's a lasting design.

Author: You consider the current Eldorado to be "a lasting design."
DR: Certainly. Look at the Lincoln Mark VIII. It's swooshy and contemporary, but its design abandons all the classic heritage of Lincoln. There were some wonderful Lincolns. The Mark VIII is very trendy, but in another two or three years who's going to care about it? That's why I think the Eldorado is going to live right through all that. The Eldorado was really designed to have a character that would evolve after the round cars. So if you look at the Eldorado carefully, it's got this wonderful shape—but it isn't round. The "crease-in-the-pants," the sharpness, all that goes back to Cadillac heritage.

But you know, we had no one in the studio at the time we did this Eldorado that had ever worked in Cadillac Studio before. The one person who really helped glue it all together was Dave Holls [a GM Design Staff executive who had worked on several Eldorados since the original '53 and had an extensive amount of experience in Cadillac Studio]. He'd come in every day and review things, or he'd bring in some pictures, maybe something they'd done in the '50s, or whatever, and everybody would look at it. In the end, this all had an effect. This helped us understand what the cars were trying to be. After we did all these models, we really began to understand.

In the beginning, it took about two months before a sketch appeared that could have been called a Cadillac. Then after about three or four weeks, Dave Holls came in

At first encounter with the 1992 Cadillac Eldorado, one is immediately impressed with the simple elegance of the design and the lack of signage and chromed gee-gaws. This coupe is graceful and sporty, whether at speed or just sitting still. When the owner of this particular example purchased it, he was impressed that "Cadillac" wasn't spelled out in big letters all over the car. The very presence of the car's design says Cadillac. Eleven inches longer than the previous Eldorado, improved braking, a retuned Computer Command Ride system, and power from Cadillac's exclusive 4.9-liter transverse mounted V-8 engine with 4T60-E transmission made this a world-class contender in the luxury automobile market. One of the concerns people had with the previous Eldorado was that the full-door treatment was too much like a sedan. People preferred the "glass into roof" design of this new car. GM Designer, Dick Ruzzin, said, "it just demonstrates how different the direction was in designing the Eldorado compared to the Seville."

The 1992 Touring Coupe featured additional refinements in the chassis and an exclusive leather interior with Zebrano wood inserts. The words, "Touring Coupe," appear on the door moldings, and the sail panel Cadillac crests are deleted. (Incidentally, the Eldorado's exterior was designed to be completely different from the four-door luxury Seville. Only one exterior panel, the sharp-raked windshield, is shared between the two models. Dick Ruzzin, chief designer at Cadillac studio at the time, said, "When we were doing the Seville, we nicknamed it 'the greyhound' because of the rounded, muscular look. We called the Eldorado 'the needle' because it's sharper and really goes off in a different direction.") The Eldorado and the Seville were built to be of the highest quality. Owner: James R. Barr, M.D., Memphis, Tennessee. *Photo by Carol Jacks*

and looked at our board of sketches and said, "You guys don't really realize it, but this is the most significant board of Cadillacs that I have seen here in ten years." He was right. We didn't realize it. We were just trying to start a new Cadillac design culture.

From that pool of work, we just kept evolving using the information from the product clinics which gave us "the voice of the customer". The customers said they wanted the car to be distinctive. They wanted it to be American, clearly American, for the Eldorado and international for the Seville, but in the case of the Seville, an American expression of an international sports sedan.

This is what we tried to do. Although much of our work was very calculated, a great deal was emotionally driven. The emotional side had many facets, including our design group's enthusiasm for Cadillac, for cars of all kinds. One emotion we totally underestimated was that of the American buying public, who reacted so strongly, with great enthusiasm, to a truly American car that could be rated with the best from Germany and Japan. It was expected that Cadillac should meet the challenge, and they did, but the fact that it was done in such a clear, decisive way was a credit to the entire Cadillac team—engineering, marketing, advertising, sales, and design—of which I was very proud to be a part.

Leonard Casillo Interview
Downsized Cadillac, the current Eldorado and Seville, and The Future of Eldorado

Leonard Casillo graduated from the University of Bridgeport, a Connecticut engineering design college. He joined General Motors in 1961 and first worked at Cadillac Studio as Stan Wilen's assistant from the late 1960s to the early 1970s. Casillo became an executive at Design Staff overseeing Buick, Oldsmobile, and Cadillac (BOC), spending most of his time with Oldsmobile. He was reintroduced to Cadillac at the end of the downsized Cadillac model run and worked on the current Eldorado and Seville.

Author: As I understand it, when you first were at Cadillac, you worked with Stan Wilen on some facelifts of some of the early-seventies Eldorados. As an executive of BOC, you later returned to the Cadillac programs in time to work on the facelifts of the downsized Eldorado and the Eldorado currently on the road.
LC: Yes, we did some minor facelift work on the ill-conceived downsized Eldorado. We attempted to staunch the flow because those cars were never really all that well accepted. We did some patching and some work to try to add a little length to the car. The last year or two before the old bodystyle of the Eldorado bowed out, it had been the recipient of a number of changes designed to give the car a little bit more presence and stance in the marketplace. We added some shape to the hood, extended the front fender profiles, and added a little bit of planing view to the rear quarter. This was all designed to make the car look a little bit more impressive and to give it some presence.

Author: The idea was to help the sales of the downsized Eldorado by doing a careful facelift?
LC: Yes, I think most of the work was cosmetic in nature: for example, creating a fender peak where there wasn't a fender peak, which added visual length to the car. We thought this added a certain formality to the car. A lot of us who had worked on the original car thought formality was lacking in the original downsized version.

Author: Let's go back to the first ones, the late sixties. Were you working mostly on facelifts at that time?
LC: Yes. I don't remember Stan and I together ever working on a major design for the Eldorado back then. I'm sure there were a number of facelifts that we worked on together. I know I never worked on the car that preceded the downsized car (the 1979 Eldorado). I think that car was probably done by Wayne Kady, who works for me now in Buick. I think Wayne at the time was the chief assistant when they did that particular car. That was the one that many of us considered to be the classic Eldorado, the one that preceded the downsized car.

Leonard Casillo.

Author: So as far as the '71 is concerned, you don't have any specific memories of working on the development of that one?
LC: No, I think anything I had to do with that body would have probably been in some of the subsequent facelift years. My memory gets pretty foggy. I'd have to actually sit down and go back and look, but I don't remember the usual kind of excitement that's generated when you're working on an all-new car like that. I don't seem to recall ever having played that game, although I know we did work on a Coupe DeVille and a new Sedan and I seem to think that was the car, when Stan and I were together, that we spent the most time on.

Author: Okay, so let's move on to the downsized one. You were telling me about some of the cosmetic things that you worked on.
LC: I think we probably would have to go back a couple of years before the introduction of the downsized Eldorado and Seville. We put some fairly late-breaking changes into that car to try to keep it alive. We had already started thinking about what the next-generation Eldorado and Seville would be, and we knew that we had to recapture a lot of what was lost from the reputation of the car that preceded it.

At the same time we were still doing things to the downsized car to try and give it a little bit more life in the marketplace. We realized that the new car was still a number of years away.

The type of thing we did was add a little bit of shape: change the line of the roof or add a power dome to the hood. All this gave the car a more powerful and impressive form.

And we created what looked like a little bit of a fender blade that came off the number one pillar in through the front fender. This put a certain formality back in the car and tended to add just a little bit of visual length to the front fender. Our changes made the whole profile look a little bit more imposing than the original downsized car, which had kind of a truncated look.

I don't remember if it was all in the same year, whether it was the year that preceded the front-end change, or whether both of those changes were kind of lumped into one package. I do know these changes were made during the general time frame that we're talking about, in the last few years of the life of that vehicle. We did add a standoff taillight to the back of that car where there was almost a fender blade coming down the rear quarter from the sail panel. This was a taillight that you can view now in side view as well as in rear view. The original lamp on that car was sort of a flush-mounted affair, and there were no quarter extensions or fender peak on the car.

The original downsized design was just kind of milled off and you couldn't see much of the taillight in the side view. The car again lacked any real stature when you viewed it either dead on, or, especially from the side. Our changes were designed to add a little visual length to the end of the car. I'd have to go back and look to see whether or not we had to add any real physical length to the car. I'm thinking we were able to facelift the car and still keep the bumper at essentially the same location.

Author: I know Irv Rybicki wasn't very happy with the whole downsized program, of course, either.
LC: Right. In fact, I remember Irv at the time, fighting the downsizing directive almost for the entire program. There was a very strong mission statement from within Cadillac Division to downsize that car at any cost. Better-to-be-too-small-than-to-be-too-large was kind of the mentality that the division had at the time. There was such paranoia about future skyrocketing fuel costs and the need to be economical. Weight had been directly tied to fuel economy, so every quarter inch of length was fought for as holy ground. Cadillac Division very clearly said they would rather err on the side of smallness than to get out there with another car that was too big, and therefore perceived as inefficient.

Author: So they did.
LC: Yes. A lot of us felt, even during the development of the car, that this was not going to bode well for us. We could tell that, while we might be meeting the mass and weight goals, we were doing it at a terrible cost to the image of Cadillac, especially from a styling standpoint. All I can say is that the designers did the best we could all during the design process of that vehicle.

I did not work on the original downsized Eldorado. In fact, I was involved with the Toronado which was a sister vehicle to the new Eldorado. We didn't feel quite as bad about the Toronado because it was a rather different design. It was not the flagship for Cadillac and didn't need to do a lot of the things that we thought the Cadillac had to do.

I remember listening to Irv as we'd get together during lunch and talk about where we were going. There was a tremendous concern that the Eldorado was not accepting, or the Seville was not accepting these new architectural requirements well. Toronado was doing a better job because it was quite a departure from what Eldorado had to be. It didn't have nearly the heritage to bring forward, which we thought the Cadillac did. Even at that, we were almost shutting the door on the Toronado's path and striking out in a new direction.

I still think that there were certain aspects of the downsized Toronado that weren't that bad. Of the three, the Riviera, the Toronado, and the Eldorado, I think the Toronado maybe fared the best.

In many ways, the Eldorado suffered the most because there was such a predetermined sense of what that car should be. In the marketplace, there was just a tremendous disappointment when the car came out looking so small, kind of an underachiever. It just didn't have the kind of look that you would attach to a vehicle with those aspirations.

Author: Was this before the days of Grettenberger?
LC: I'm trying to remember now. Was there another general manager in the very beginning? Yeah, it was probably right on the cusp. My inclination is to say that John inherited a lot of that and I'm trying to think who his predecessor was. Ed Kennard and Bob Berger were the general managers jointly responsible for the '86 development.

Author: I just can't imagine that coming from Grettenberger.
LC: My perception is when I came back on board at Cadillac, we had a year or two of trying to patch those downsized cars together. I had this sense that a lot of what was out on the floor was stuff that John Grettenberger had inherited. He was such a strong supporter of the new car (the current Eldorado) when we started it, that I've got to believe that even if he had been a part of that downsizing, it would have been as an unwilling participant and that the orders probably would have come from above him.

I think there was a very definite mind-set all the way up to the top floor in the corporation that we needed to keep the cars that were more fuel efficient. Cadillac, I remember, was sensitive to the times that we were going through, because they had the image of having the largest cars in Detroit—the most fuel inefficient cars. Cadillac was going through this kind of thinking that, if the hammer ever falls and fuel does go through the ceiling, we're in trouble. If the American public becomes obsessed with fuel economy and finds it socially unacceptable to drive something that looks inefficient, even if it isn't that inefficient, we're in trouble.

We were very concerned about Cadillac because its physical size had always been a strength in the past. In fact, you don't have to go back too many years where you would argue that advertising should have touted things like the size of the car—the bigness of the car. Now all of a sudden, there was this growing unease that what had been a strength could become a liability. Suddenly, there was this big rush to overcompensate within Cadillac.

Even those of us who were concerned about the look of a car were also worried. We couldn't ignore the fact that the corporation was thinking that strongly about this problem. I remember myself, just a designer, wondering about the economic thinking that was going on in the bowels of the corporation. I remember thinking, "Boy, this must really be serious because there's so much concern that if we don't get these Cadillacs down to size, we're going to have a hard time selling them in the marketplace."

Downsizing wasn't even something that you questioned because it was such a compelling concern on the part of everybody that was working on cars for General Motors. You just didn't question that. Downsizing was inevitable. The word from the very top minds of the corporation was that we had to get our cars right with what seemed to be the coming trends.

At the same time, as designers, we could see what this was doing to the cars, especially at Cadillac. I can remember saying, "Boy, I hope the world is ready for this." Unfortunately or fortunately, none of the things that we all feared were going to happen happened with the intensity that we thought they would. Had they happened that way, had fuel gone as high and become as scarce as some of the doomsayers had predicted, those downsized cars could have been very successful cars.

As it was, because we, in effect, abandoned that very traditional segment of the market where big is good. Ford took a different tact: they just made the decision to keep building some of their old cars. For them, these became cash cows for the corporation and they became the only game in town. The old Lincoln Town Car was nothing to write home about. It was an old car even on Ford's books, but they made the decision. Maybe they had no choice, maybe they didn't have the money to invest in retooling and downsizing the car. So they just kept their fingers crossed and put it out there. And, lo and behold, the price of gasoline didn't ever go where people thought it was going to go. It rose some, but it wasn't the end of the world and the end of the availability of gasoline as a fuel source. So with Ford's cars, all of a sudden quality crept back in the picture. If you wanted a great big car that could haul six people around—and all of their luggage—that was the only game in town! It made Ford very wealthy during those years. Meanwhile, Cadillac had cars that were prepared for a world that never quite arrived.

Author: Someone had told me that there was a meeting where it was seriously considered to have a three-cylinder Cadillac. Had you heard that?
LC: No. Dick Ruzzin would know. Wayne Kady was in Cadillac when I came in as an executive. Less than a year later, we moved Wayne down to Buick and brought Ruzzin in. Then Dick stayed on after I left, but that wouldn't have been something that would have been discussed while Dick was in Cadillac.

Back then we were developing the '89 DeVille, which was kind of the Band-Aid to the DeVille, and which, while it was suffering some, didn't suffer as badly as the Eldorado and Seville. That was kind of a rush program—to hold the wolves at bay. We were wildly successful with that car, way beyond our expectations. That gave us the time to do the job right on the car that just came out this past fall, the new Concours

Cadillac Allanté, body designed and built in Turin, Italy, by Pininfarina. As part of Cadillac's commitment to restoring the Eldorado to its former glory, every effort was made to ensure that the best design for the car would be developed in the Cadillac Studio. Pininfarina, because of their long association with General Motors, was asked to prepare concept designs for the process that would be considered right along with the designs being developed at Design Staff. Because of the complexity of the design process, especially in this case, it is impossible to say how much of the final Eldorado design evolved from Pininfarina's contributions. The Chief Designer of the Cadillac Studio at the time, Dick Ruzzin—the man who guided the entire process, could not say when we spoke with him, and neither could Sergio Pininfarina. However, several designers suggested that we look at the front end of the Cadillac Allanté which was designed by Pininfarina, in an effort to assess the influence of the Italian organization on the 1992 Eldorado.

and DeVille. These cars represent the final evolution of that bodystyle. Dick was largely responsible for the new line.

Author: What about the current Eldorado? What do you know about that program?
LC: The one that's on the road right now? Well, that's the program that I was most heavily involved in with Dick Ruzzin, both the Eldorado and the Seville. That was the program where we finally had the—I won't call it a blank check, but clearly Design had been given a mandate: "Guys, we've got to go back out there and we're not going to go crazy, we're not going to give up all the gains that we've made, but if a few inches of length gives you the opportunity to do a better design, we're not going to absolutely, stubbornly refuse to consider that. We don't want to go backwards to create a behemoth again, but we do want to make styling a major win on the new Cadillac."

We said, "Now we clearly have the mandate that styling has to win, that we have to have a success in that arena." We would do what we had to do to make sure that we met our fuel economy targets and we would meet our mass targets, but we wouldn't do this at the expense of styling. So right from the git-go on that program, we knew that we were going to be looking at a significant re-proportioning of the car.

The beauty of that program was that there were a couple of things in our favor. The Cadillac organization now had fully matured as an organization that had its own manufacturing and engineering group within its confines, as opposed to the organization that we would have in place today, and as opposed to the organization that preceded that. Cadillac was one of the first of the divisions to have, in effect, its own dedicated plant and its own dedicated engineering and manufacturing resources. So we had the whole team. It wasn't like we had to go out and sell it to the platform. Cadillac and the general manager were able to sit down with all of the resources necessary for that car at manufacturing end, engineering, and design, and as a team call everybody together and say, "OK, guys, we got to do a car now that wins on all these fronts and we won't win in any one area to the exclusion of the other. We've got to have an exceptionally good looking car. We've got to have a good performing car, good quality car."

We forged a team that was comprised of the engineering and manufacturing people. It was a fairly unique experience because it wasn't a typical situation where we, as the design end of the business, would deal with the marketing division and the marketing division would call on the engineering end. Then at some point, we'd start dealing with engineering and at the very end of the program, manufacturing would get on. The entire operation was in lock step and it evolved from day one.

In the earliest reviews of clay models, the manufacturing and engineering folks were in the building. We all knew what the goals were so there was no arguing; there was no infighting at all about what had to happen. Everybody knew what had to happen. And, the folks who were responsible for the mass targets knew that they couldn't take it out of our hide. They knew that together we had to satisfy their mass targets. We had to do what we had to do to help them, but at the same time they couldn't just walk in and take six inches off the front end of the car and say, "Sorry guys, you make it look the best you can but we can't afford to do that."

It was a tremendous team effort. I remember that. It was an exciting, exhilarating period where the group that came together kind of had a mission and a purpose. We realized that we only had one more time up at bat. I don't know how many times we said that, but we all said that.

We recognized the downsized car we had was not successful, and that we couldn't go out there with another car that was just a little successful or almost successful or even a little bit better. We had to knock one out of the park, or the marque was probably not going to survive to another facelift.

Author: So you think the division was at stake?
LC: Well, if not the division, certainly the marque was at stake. If you look at what Seville and Eldorado have done for Cadillac in terms of the umbrella effect, I think clearly that without those vehicles having gone out there and capturing the sympathies of the writing press, the automotive press, I don't think we could have staged the comeback that we have. The cutting-edge presence of Seville and Eldorado made the difference in the marketplace.

It was Northstar, it was drivetrain, it was proof quality, dedicated hand-tram and pull-down assembly plant, it was the look of the car. So there were a number of factors that were all engaged to make that car work.

I think the success of the Seville and the Northstar are what gave us momentum on the new DeVille and Concours, and it really kind of, in my mind, wrapped up Cadillac's equity in the marketplace—a significant amount over what Cadillac had prior to the introduction of those cars. So whether the division was at stake, that would be kind of a hairy statement for a designer to make, but I certainly think the comeback and the success that we are currently enjoying would have been severely curtailed if not for the success of those vehicles.

We clearly knew it was very important that we weren't going to get another chance to do that car. If that car hadn't done well, we probably would be shrinking the division. At the very least, we'd be looking at maybe putting all of our eggs in the full-size luxury passenger sedan. Certainly, we'd be abandoning the sports sedan and the sport coupe end of the market if it weren't for the success of the current Eldorado and Seville.

Author: Now, in developing the Eldorado, as I understand it, there were three models that were started simultaneously. Were you involved in that part of the process?
LC: Yes, probably we generated more than just three models, but we had three teams working on it and that's the three that I remember. We had Pininfarina and his operation involved in working on that vehicle from the very outset. They generated a fiberglass model and a number of scale models prior to it. We were working on a model in the Advanced Studio. We did a completely separate series of scale models that led up to that and a number of variations on the full-size clay that we did downstairs, and Dick Ruzzin and his folks did another car up in the production studio.

So, there were three full-clay themes at any one time. But we did a total of several variations among us. I know several fiberglass models were done. I know that Pininfarina did one. Later, he made some modifications and resubmitted that car. This effort was a full-court press and we used all the resources in our building we could muster—in addition to the Pininfarina operation.

The Italians were flying back and forth on a regular basis. This was a friendly competition. While each group would have obviously liked to have done the car, we didn't hide anything. It wasn't a question of, "Well, here comes Pininfarina, let's cover up everything!" No, we opened the doors wide and when Pininfarina and his design team were there we kept them involved in everything we were doing, kept them abreast of everything that we were doing. They showed us everything they were doing. We even showed them all the advanced things we were working on. It was a friendly competition, but at the same time there was a bigger goal that went beyond anybody's individual success: that was that we had to do a great car!

I think Pininfarina was flattered that we asked him to participate. Given his long association with Cadillac over the years, I know that he took it as a very, very serious assignment. It was almost a solemn thing with him that his team did the very best they could to support the project.

The beauty of that competitive mix was that I cannot tell you right now whose car "rang the bell." Maybe the car that Ruzzin and I did in the production studio became the ultimate car, but that's simply because that studio was charged with taking the chosen theme into production. Sergio Pininfarina and his team had a very significant effect on the design of the car that we ended up with. The Italians affected the way we thought, they affected the way we designed. We all recognize the fact that, if it weren't for their involvement, the car would have looked differently. The Advanced Studio also played an extremely significant role. Ruzzin's studio was the place where we simply brought it all together. There were good things from all three teams that were all brought together in the final design.

Author: Was this team approach used because of Grettenberger or because this is the way things are done at Cadillac now?
LC: Cadillac was always unique. Saturn is like that, too. They both "own" their own resources to shape their own destiny, so to speak. I always thought this was all part of John Grettenberger's vision. He sort of had the assignment to bring Cadillac back on track. He had inherited a lot of the consequences from what had been put on the road during that downsized period. I am not sure if he used this analogy or not, but I know some of us did. In some of our meetings, he emphasized the idea that we have all of the legs of the stool—engineering, manufacturing, and design—working together to support the creation of this new vehicle and Cadillac's reemergence in the marketplace. I attribute a lot of this cohesive working together to John Grettenberger. I think he thought he needed that to do what he had to do with Cadillac Division.

Author: What do you think is the future of Eldorado?
LC: Well, we have since reorganized. We now have CLCD—Cadillac Luxury Car Division—which is the group that has the manufacturing responsibility for Cadillac. Ironically, it largely has the same people in it as when Cadillac had its own manufacturing group. Only now, those people have taken on some additional responsibilities. CLCD will be the platform for a number of vehicles besides just Cadillac.

I think that what the "Camelot" of what we had when we did the current Cadillacs gave us the vision and the momentum to move on to other things.

I can't say enough good things about the new relationship Cadillac has with CLCD, the platform division. So even though we don't "own" that division as when we did the current models, the fact that most of them are Cadillac people, everybody has a sense of mission now.

I'm not afraid that, just because Cadillac is not like Saturn anymore, that we'll lose our way like we did before. CLCD in Flint is peppered with Cadillac people. The loyalties and interest are still there.

I can't talk about Eldorado without talking about Seville. Those two cars have carried a new message forward.

There was always the feeling that the Eldorado would appeal more to the American sport coupe buyer and the Seville would appeal more to the import sedan buyer. Large luxury coupes seem to still be more of an American phenomenon than a European phenomenon. Sporty sedans that look and feel like coupes seem to be more of a European phenomenon than an American phenomenon.

Our two-prong approach is that the Seville would do some conquering of import buyers and the Eldorado would pick up the flag for the American ideal of the personal luxury coupe. I don't see that changing significantly. The only thing I'd say is that those two approaches may be converging. In some later generation there may not be so much difference between them, except to say that one is a two-door and one is a four-door.

We are just now starting to think about what the next generation will be. We certainly haven't decided what we want to do. We've been conducting a lot of research right now, primarily listening to the customers, probing for possible weaknesses in the current vehicle, and what would make the current vehicle stronger. I get a sense from that that the two bodystyles may become more closely intertwined—at least from the standpoint of one being domestically oriented and one being European-oriented.

To explain something, I have to give you some background: we did the Eldorado before we did the Seville. We had a lot of upfront energy on the Eldorado and we came in on the Seville only after we had sort of set the direction for the Eldorado. The Seville took an entirely different approach. It was one of those things where we got lucky quickly. Where the Eldorado had been a very long soul-searching exploration, the Seville just happened!

In fact, we were so pleased with the Seville's roof line that we went back and did a proposal for the Eldorado with that roof line. One of the Eldorado proposals looked very much like the two-door Seville that was in the auto show a few years ago. I think that the reason we didn't pursue that was the feeling at that time that those two cars had to present two different faces in the marketplace. And, there was the feeling that the Eldorado was going to appeal to the more traditional segment of the market. This meant we had to keep that formal heavy sail panel in because everyone knew that the heavy sail was a very acceptable look among the traditionalists.

Actually, there is a strong family resemblance between the two cars: body sizes are very similar; it's in the rear end and roof lines that we really get the separation between the two models. The Seville is more open and the Eldorado is more closed.

Marv Fisher Interview
Interior Design of the 1992 Eldorado

Marv Fisher joined GM in October of 1967. Since 1980, he had been the assistant chief designer at Pontiac Studio, where he had worked on the Pontiac Fiero interior in the concept studio before going into the production studio there. He was responsible for much of the change in Pontiac interiors during this period. At the suggestion of Dave Holls, Fisher was assigned to the Cadillac Interior Studio as chief designer in 1987 to work on the all-new generation of Cadillacs.

Author: When you arrived at Cadillac Studio, what was the status of the new Eldorado interior?
MF: The new Eldorado had not been started at that time. They had a theme and an advanced expression type of thing that didn't end up being in the final design anyway. Drew Hare, the chief interior designer I was replacing, worked with us for about three months before he retired. While he was there, we did start on the interior and he was involved in the early stages, but the theme changed quite a bit after he left.

Author: What was the overall concept for the interiors of the Eldorado and the Seville?
MF: You have to understand that it was a new team that took over in the Cadillac Studio, both for the interior and exterior. For exterior, Wayne Kady moved into the Buick Studio after a couple of months, and Dick Ruzzin came in from Chevrolet. So the exterior group was a new team.

My predecessor, Drew Hare, had a definite philosophy of design. He had worked on the larger Cadillacs and then the downsized ones of the mid-'80s. His philosophy was one that included forms that were very crisp, very formal, square, and so forth. All the woodgraining would have beading and whatnot—he took this concept right through the last Cadillac interiors he did. This, combined with the downsizing, made for an unfortunate circumstance during that time.

For the generation of Cadillacs my team was to work on, we leaned toward a more international flavor. At this point, the interiors of Seville and Eldorado began to take on more of a sports/luxury/personalized style as well. So there is definitely a sports flavor to the international direction.

As far as form and function were concerned, my background at Pontiac had been more of a flowing, organic type of thing. The exteriors were already headed that way, and I always tried to make it a point to have the interiors match the exteriors—giving an overall harmony of form. So you'll see that the form of the instrument panel is shaped more like the hood of the car.

Some designs overstate the round shapes and forms. Such designs might apply to niche markets or be used as something fun and expressive in a low-end market situation. Cadillac, because of its refinement and elegance, requires some sort of control over those flowing lines. You'll get some straighter lines as the flow is generated. In other words, we tried to add some finesse to the forms.

Marv Fisher.

We also tried to pull in traditional elements to blend with this new stuff, so you'll find things like woodgrains in these new interiors. We chose real wood for use in both the base and up-level models, but even the shape of the wood reflects a different design philosophy from what was used before. This isn't the square-edged, chamfered stuff. It actually takes on form instead of being just a flat surface.

Later in the program we decided to use round instrument gauges. This was a strong international cue. Other elements of the controls are unique to each division. For instance, Cadillac handles climate control and the driver information center differently from how the other divisions do it. These were all elements that drove the architecture of the interior.

There was a controversy at the time. The traditional people wanted a three-across seating arrangement, kind of an open cockpit, whereas the people who preferred the international flavor wanted the full-console environment. We tried not to compromise the concept by going from one scenario to the other, but these ideas do drive certain types of designers. That's why the heater control is not down by the radio, but next to the instrument cluster. Doing this saved us from moving so many components from one area to another.

On the right side, for the first time there was a passenger's side airbag system that had to be tested so it would comply with regulations. GM had worked with airbags for a number of years. The units had always been a bulky and massive item on the right-hand side, but Cadillac did a real nice job reducing the size. This made it easier for us to do our job, and gave the customer more space on the passenger's side.

Author: For this project, you worked closely with the exterior designers.

1993 Cadillac Eldorado, production car. Outside rear-view mirrors now match the exterior color of the car. Two new exterior colors were added to the list, Light Beige Metallic and Dark Cherry Red Metallic, for a total of seventeen exterior color choices. For 1993, the Eldorados were finished in a new "waterborne" paint process. Eight interior colors were available, and dual airbags were standard. Knee bolsters were provided for both driver and passenger to prevent "submarining." An express-open sunroof was now available as an option.

MF: Yes, the studios were right next to each other. We had the same entrance and there was just a divider. As we have worked more closely together as teams, they have tried to pull these studios even closer together. Cadillac was itself drawing itself together, almost like its own complete car company. They were setting up car teams. We had one manager. Cadillac involved us designers in their meetings, their drives, and their handling reviews in order to get that kind of continuity in place.

Author: Because the new cars in the Cadillac line have similar interiors, does this mean the division is returning to the time when there was a much greater similarity between the various models?
MF: There are quite a few elements operating there. Cadillac was trying to do as many of the models as possible at their Hamtramck [Michigan] Plant. Once you decide to do something like that, you can benefit from similarities between the models. For example, once you have solved the problem of how to handle the airbag restraint system on the right-hand side, then you'd prefer not to have to redesign it again for the next model. So some of the basic architecture begins to set up, pulling the various models closer together. It makes the system more efficient.

However, there were some luxury cues that were in the Seville that were not in the more sporty Eldorado.

Author: Did you purposely limit the number of colors used in the Eldorado interior?
MF: We wanted to get good colors. We thought that to do this, it would be best to limit the proliferation of colors. There wasn't a real strong effort to limit the number of colors so much as it was to get the best colors possible. Proliferation is a driver for lack of efficiency, so choosing this way of handling color was more efficient.

In fact, there are creativity teams working in GM to handle this sort of thing. When all these platforms broke apart from a central organization to a series of more individual ones, this generated a lot of proliferation. Colors and that kind of thing are all possible areas of proliferation. Getting a handle on this problem increased profitability within the corporation.

Author: The seating in the Eldorado is impressive.
MF: This was to have the character of an international driver's car with very good lateral support. The structure of the seat was firm and rigid, with adjustable lumbar support and expressive headrest shapes. A lot of study was done to make the seating right. For the materials, we got a much softer grade of leather and took the gloss level down from what had been used before. That way it didn't look like a plastic.

In fact, there were three major material changes for these cars: the nicer grade of leather, the wood, and the carpeting. The carpeting was greatly upgraded. It was strong and durable, but with a contemporary flavor.

Author: In driving the car, what is your impression of the interior?
MF: As you're driving it, the flow of the instrument panel design and the surface of the hood complement each other in the movement of the car. This makes the turns and other maneuvers more harmonious.

The harmony of the interior was a hit at the auto shows. I remember that the Mercedes people were really impressed with it.

Robert C. Dorn Interview
The Development of the Northstar System

Robert C. Dorn joined General Motors in 1956 and graduated from GM Institute in 1961. He first worked with Cadillac from 1979 when he replaced Dan Adams in his chair (as he puts it) to 1981, when he went to Pontiac. In mid-1985, he rejoined Cadillac as chief engineer of the Detroit Product Team, which was resident in the same area as Cadillac Marketing. In 1987, when Cadillac was reorganized, he became part of Cadillac once more, as general director of Operations. A year later, he became chief engineer again. Presently Dorn is chief engineer of the Advanced Vehicle Development Center in GM North American Operations. Dorn is considered by his peers to be one of the primary forces behind the creation, development, and success of the Northstar System.

Robert L. Dorn

Author: When did plans for what would become the Northstar System actually begin?
RD: That started about 1988. We were in a co-location with the BOC (Buick, Oldsmobile, Cadillac) powertrain activity led by Tom Stevens. Tom had been with Cadillac his whole career. When we divided things up, Tom ended up being the chief engineer for Cadillac engines and powertrain. Tom and his team started to talk about what might be the next generation of Cadillac powertrain because we were working jointly to get Cadillac on the same playing field with the very best cars in the world. Our feeling was that we needed a powertrain beyond a 4.9 liter pushrod engine. Even though that was a good solid conventional 200 horsepower, it wasn't the kind of engine we would need to compete with the best luxury car companies in the world.

So Tom made a proposal, and we developed what at that time became the Northstar System. At first, we called it the Balanced Car Concept because we knew that when we were adding that degree of horsepower and performance, that we needed to do something more than to add just that powertrain to the car. Subsequently the marketing name for that system was the Northstar System. Of course, there's lots of published detailed information available on the specifics of the system.

Author: Where did the name come from?
RD: I believe it came from the guys working on the project—the original engine team. They were looking for a name for their project and came up with that. The name stuck as the project went through the division. The Cadillac marketing people did a lot of work on it and decided it was a really good name. They said it tested well and we might as well use it, instead of changing. Of course, the press by that time had nosed around and found out about the Northstar project and they were wondering what it was.

Author: What is the basic idea behind the Northstar System that sets it apart?
RD: With the brand-new engine and brand-new transmission, we knew we were adding a lot of torque and horsepower to the car. We knew that one thing Cadillac owners don't like is a car that's out of balance or surprises them. They don't like technology for technology's sake. They want to get something for the technology and, above everything else, they want it to work. They'd prefer not to get it if it isn't going to work well for them.

We were pushing the state-of-the-art for front-wheel drive in the world in adding that much displacement, that much torque, and that much horsepower driving through the front wheels. We knew if we did all this, we would have to rebalance the car. Our engineers then got together and came up with what would become the rest of the Northstar System. By that time we had ABS brakes as standard equipment. We then said we needed traction control. We knew we needed traction control that "talked" to the engine—not just hung on to the brakes. We also wanted to put a suspension system that would be better balanced, so we did an SLA rear suspension and added real-time damping (the latter would give us better control at high speeds over any kind of roads). We put larger brakes in the car. We also did some aerodynamic work with the design guys and worked on a few other details.

In the end, we had done quite a number of things to the car to rebalance it so the customer could be in control—not the car being out of control and the customer having to try to control it. We did all this because we wanted a very refined front-wheel-drive automobile, because that was what we had and that's what we wanted to go on with. We wanted to make sure the customers would like it.

John Fleming was the director of Marketing at Cadillac and he knew all this would increase the cost of building the cars and therefore increase the price to the customer. He wanted to make sure the customer wanted this before

Left, the 1993 Cadillac Eldorado Touring Coupe is instantly recognizable by the reduced amount of chrome on the front end and the lack of Cadillac medallion on the sail panel. The big news was the Northstar System, which included not only the new engine, but a fully electronic transmission, advanced Road Sensing Suspension (RSS), Speed Sensitive Steering (with Electronic Control Module, ECM), Short/Long Arm (SLA) rear suspension system, and Traction Control. This was the new Eldorado reborn yet again! Suddenly you could purchase the most sporty elegant luxury sedan in the world *and* accelerate from 0 to 60 in 7.5sec or cruise comfortably at a top speed of 150 mph. Cadillac released the following statement (August 25, 1992) on its Northstar engine and matching transmission: "Both the block and the cylinder head assembly of the Northstar engine are constructed of cast aluminum. For increased rigidity, iron bore liners are cast into the block. Despite the additional valves, camshafts, and related chain-drive components, the new engine weighs just 230.1kg—comparable to the current 4.9-liter engine. Rather than the traditional five-bearing cap assembly for the crankshaft, the Northstar engine features a two-piece block with a large, one-piece girdle assembly. Often found in high performance or racing engines, this arrangement provides an extremely rigid carrier for the crankshaft. The compression ratio of the engine is 10.3:1 (final drive ratio of the Eldorado Touring Coupe is 3.71:1). The new engine produces 295bhp at 6000rpm and 290lb-ft. of torque at 4400rpm. Engine redline is 6500. EPA-rated estimated fuel economy is 14mpg city and 21 highway. Mated to the Northstar engine is the 4T80-E transmission. Completely new, the transmission is fully electronic and actually programs itself to the driver's style. Designed specifically for the Northstar engine, the transmission features equal-length drive axles to help eliminate torque steer." In addition to the "Touring Coupe" insignia on the car, the "32V Northstar" badge was displayed on the deck lid.

Left, ASC Cadillac Eldorado Convertible Concept Vehicle. Back in the mid-1980s, before the 1986 downsizing, ASC had built GM-sanctioned Eldorado convertibles. In partnership with GM Advanced Engineering Staff, ASC built this Eldorado which features ASC's one-button top which automatically lowers all four windows, as well as the top, with the touch of a button. The exterior features a contoured, hard, tonneau cover that conceals the top when down and accentuates the grand touring look of the car. The vehicle is finished in Champagne Amethyst (a rich red-brown) and sports a color-keyed top with neutral leather interior and chrome production wheels. The glass rear window has an integral defogger and the headliner is color-keyed. ASC's vice president of design and program development said, "ASC spent many hours to maintain Cadillac's image and design features on the Eldorado Convertible concept car. We built the vehicle to explore press, dealer, and consumer perceptions on three levels: first, to test ASC's latest top technology; second, to investigate the potential want for the convertible in this market segment; and, third, to evaluate their impressions of our design execution. In addition, the overwhelming interest that the Eldorado Convertible received at the North American International Auto Show convinced us to show the car in Geneva (Switzerland)."

we offered it to them. To do this, he actually ran clinics of cars that were built up by ourselves and Powertrain. to simulate this kind of performance. The customers acknowledged that they really did want this, even though it was probably more performance than they'd normally want. But they thought it was a very important part of the safety and security of their Cadillac—even when we talked about,

"What if we raised the price of the car this much, and that much?" and so forth.

This sort of broke the log jam inside Cadillac that this was the right thing to do. And for more than just the enthusiasts, this was the right thing for Cadillac customers. After this, we had a lot less internal doubt, and we were on our way.

Right, Cadillac Eldorado, ASC-modified Eldorado built to Chuck Jordan's specifications. Before he left GM as vice president for Design, Chuck Jordan had American Sunroof Corporation modify an Eldorado to his design specifications. He discusses the car at length in his interview in this book.

1995 Cadillac Eldorado. The most radical facelift yet on the original 1992 Eldorado, the 1995 has new front and rear fascias, fog lamps, bodyside moldings, cast-aluminum wheels, three new exterior colors, and a more powerful Northstar V-8 matched to a new Integrated Chassis Control System. The Eldorado Touring Coupe version has a more aggressive look than previous years. Lean and mean, at first glance the Touring Coupe has a new body-color grille, new front and rear fascias, fog lampd, and an even more restrained use of badging and brightwork that enhances the sporting character of this splendid machine. This image is reminiscent of the special Eldorado Chuck Jordan had American Sunroof Corporation build for him a couple of years ago.

American ♦ Classics

Chapter 7

The Future of Cadillac Eldorado

By the time the front-wheel-drive 1967 Cadillac Eldorado was introduced, only 34,781 cars bearing that name had been manufactured since the model's debut in 1953. The production figures from any given year had never been impressive. Only 532 were produced in 1953 and, before 1967, the highest production figure was 6,050 in 1956.

In fact, production figures for those early years are easily swamped by the annual figures of any given year since 1967. The very first front-wheel-drive production year of 1967 put 17,930 new Eldorados on the road—more than half the number that had been produced in the car's fourteen-year history! The number of "last convertibles" built in 1976 alone totaled 14,000.

We mention these figures to give a quick overview of how the expectations of Eldorado production have changed since the inception of the original concept of the "golden ones"—Cadillac Division's opulent, luxurious cars. For over a decade they had been a truly distinctive car (amounting to only 4 percent of total production even during the high-point 1956 model year). By 1984, Eldorado accounted for 77,806 cars out of Cadillac's production, this figure accounting for more than one third the number of *all* Cadillacs built in 1967. In other words, Eldorado has become not only a special, classic automobile, but also a highly significant part of Cadillac's total annual sales.

As several people interviewed for this book pointed out, the two-door coupe is now at a low point in its popularity. The four-door sports sedan of European proportions somehow has become more popular than the traditional American two-door coupe. This trend may or may not last.

Recently, Cadillac said that it was discontinuing its two-door Coupe DeVille, an automobile that had become almost an American icon in its prestige and popularity. Several people we spoke to at General Motors said they wished such

Cadillac Eldorado future concept rendering by designer Larry Erickson. The reader should not for a minute believe that any of these renderings of possible futures for Cadillac Eldorado will ever appear on the road. These are two-dimensional *ideas* from some of the master designers working in Cadillac Studio. Some components of the designs may work their way through to production, but here we are dealing with preliminary concepts. Erickson's idea is interesting in the way it manages to include several Eldorado design cues, including the horizontal front grille and, judging from this side view, the vertical taillights. This futuristic coupe looks fast, comfortable, sporty, and elegant.

formal announcement had not been made. Many think the Cadillac Coupe DeVille will be back. Perhaps.

We must remember the "discontinuing" of the Cadillac convertible—the last Eldorado convertible in the mid-1970s–and how it underwent a forced resurrection a few years later.

To be brutally honest, the four-door Seville is now doing much better than the two-door Eldorado, both in sales and among automotive critics. But, to be fair, one has to consider that Seville is really a part of the Eldorado tradition and concept. In the fifties, the Seville name was used to designate the hardtop Eldorado from the convertible Biarritz. When the popular 1976 Seville was introduced in April 1975, someone or some committee somewhere in Cadillac or GM decided not to reaffirm Seville's connection to Eldorado. This was, at least in part, due to the fact that the 1967 Eldorado coupe had not been given a Seville name because no convertible was made available–not to mention the fact that for years previous to 1967, no hardtop Eldorado had been available. Thus, Seville became a separate model in 1975. However, it was still a special Cadillac in the tradition of Eldorado, and still is.

We think that Cadillac is in a win/win situation with Eldorado. No matter how few or how many are produced and sold each year, Cadillac management can always point to this car and say it is their flagship; that they have a heritage of leading the way with it, no matter the numbers, and that the car is respected and admired around the world as being a wonderful example of American design excellence. The division can try new ideas with the Eldorado that would be impossible to try in their other lines of cars. In this regard, Cadillac has left in the dust Lincoln's Mark VIII and whatever coupe Chrysler might want to put into the fray. Foreign competition cannot even *enter* the field against Eldorado. Being American is an integral component of its character.

A big part of the reason for Eldorado's re-emergence is that a man who really enjoys cars and has a feeling for when something is right is the general manager of Cadillac Motor Division. John Grettenberger has essentially taken Cadillac from the rock-bottom position it held in the days of downsizing, to its renaissance of today. He has managed to put together the right people at the right time. This has given both Cadillac and Eldorado a bright future.

We look forward to fresh surprises under the Cadillac Eldorado nameplate.

Dennis Little Commentary

Dennis Little is the current chief designer, Cadillac Studio. He worked as Dick Ruzzin's assistant when the 1992 Eldorado and Seville designs were done. Little took Ruzzin's place as chief designer when Ruzzin went to Germany. Little has moved around through the various studios at Design Center. While at Oldsmobile, he was one of the primary designers of the recently introduced Aurora. Including both tenures in the studio, Little has worked in Cadillac design for about five years.

"The future Eldorado might decrease in size so that it becomes more of a personal small size coupe. Probably it will end up in a sportier, international direction. The car may get as small as it was in the downsized years, because some of the feedback we get from people is that maybe the car is just a hair too big. We'll be looking for ways to downsize the outside of the car while increasing the inside. So, hopefully, you'll have your cake and eat it too—with plenty of room inside, but compact outside.

Probably, the car will have less chrome and certainly not be as decorative as Cadillacs used to be. The car will be sleek looking, but not so round that it's a 'jelly bean' type of car. But there will *not* be a repeat of the design mistakes of the downsized years.

The car will still carry many Eldorado design cues. The current Eldorado is right for today's market, but the future one will probably not be as traditional. For the 1992 design, we tried to pull many of the traditional cues for Eldorado and put them into that car—which we didn't do for the Seville. Of the two, the Seville ended up being a little bit more successful—so, we may be stepping away from what would be traditional, when we design the Eldorado of the future.

Dennis Little, current chief designer, Cadillac Studio.

JOHN GRETTENBERGER INTERVIEW
The Current Eldorado and the Future of Eldorado

John Grettenberger became general manager of Cadillac Motor Car Division and vice president of General Motors in 1984. A graduate of the University of Michigan and M.I.T., he has worked for GM since 1963 both in the U.S. and overseas. Because of the lead time necessary for production, the downsized 1986 Cadillacs were already on-line when he assumed the top management position at Cadillac. In 1991, Grettenberger was inducted into the Automotive Hall of Fame as its "Industry Leader of the Year" largely because of his insistence on increasing Cadillac customer satisfaction and his genuine appreciation of Cadillac employees. Under Grettenberger's leadership, Cadillac is rapidly regaining its leadership and prestige in the automotive world. President Bush presented the celebrated Malcolm Baldrige National Quality Award to Grettenberger and Cadillac in 1990. In 1992 alone, Cadillac received several major industry awards including: Motor Trend's Car of the Year, Automobile's Automobile of the Year, and a place on the list of Car and Driver's "Ten Best Automobiles."

John Grettenberger, general manager and vice president of Cadillac.

Author: What has been your general idea in developing the current line of Cadillacs, including the Eldorado?

JG: This line has been quite a departure from what Cadillac has been known for "traditionally." The current model was, of course, introduced in '92. The research we did for the concept, the initial design, and the engineering were begun back in '88. Of course, some of the concepts go back beyond that.

What we wanted to do was to use Eldorado and Seville as the two lead horses in the line-up to turn Cadillac's image around. It had been a conservative division known only for its large and very comfortable cars. Cadillac had not been known for its fun-to-drive, fuel-efficient, modern, youthful-type vehicles.

Both the Eldorado and the Seville have done a great deal for us in terms of changing the average age, affluency, and education level of our customers. We've actually lowered the age by a nice margin, versus the DeVille and Fleetwood cars. The new buyers have turned out to be even more affluent than those we were selling to in the past, and, in most cases, better educated.

Author: You consider these two cars your flagships?

JG: Certainly, we do. They're very expensive, as you know, in relation to a traditional car such as the DeVille, for instance, which is our basic entry-level Cadillac. They represent the technology leadership of Cadillac Motor Car Division. The Northstar System was introduced on those vehicles first, before any others. A lot of our suspension innovations have been pioneered on Eldorado and Seville, and, certainly, many of the things we now offer automatically, such as ABS brake systems, were introduced on these cars.

Author: Do you feel that you were brought to restore the prestige that Cadillac enjoyed before the downsizing of the mid-'80s?

JG: I don't think that I was brought in to do anything but to improve the overall success of the division. We did have a lot of quality problems in the early and mid-'80s. We did have an image that was stodgy in terms of where the market was going. People were dropping out of the American luxury market to buy imports. We had a lot to do. The Eldorado and Seville have played a major role in changing things.

Author: Are you happy with how the motoring press has portrayed the new cars in its publications?

JG: I am. Of course, I'm not happy with everything that's written. Much of what is written is well founded and well established in its credibility, some is complimentary, and some seems to take a lot of license. But you have to understand that this is the way things are in this business, whether you agree with what's written, or not. If you do your job right in determining what the people in the market are looking for, and make certain that the engineering of the vehicle is properly packaged before the car reaches production, you generally get the kind of marks from the automotive writers that you deserve.

We've certainly gotten some good marks on the Eldorado and Seville. When we took the Northstar versions of those cars to Europe last year, the international press was dumbfounded. They had not thought Cadillac could build such excellent cars.

Author: Has Pininfarina's involvement with Cadillac since the Allanté enhanced Cadillac's image in the international marketplace?

JG: I think so. You can see the family resemblance in the front end of the Eldorado, if you compare it to the Pininfarina Allanté. I think that to have a designer and coachbuilder of that importance and magnitude has been good for us. It's certainly not something that just happened. We did a lot of things with Pininfarina long before the Allanté. They've had a long history of working with Cadillac that's been good for them and good for us.

Author: What do you think is the future of Eldorado?
JG: Currently, the Eldorado is the only coupe that we have in our line-up. We consider that a very important role for it to play. If you've watched the market trends, however, the large luxury American coupe market has greatly reduced in its overall size. You've got the Mark, the Eldorado, and now the new Riviera coming in. Essentially, those are the only three entries in that field. It's going to be a real fight for the three as the market continues to trim down, as more and more people go into sedans.

The Seville, for instance, is a hell of a good-looking car. It's not the least bit traditional in its approach to the buyer. Anyone who purchases one will be considered a youthful and aggressive-type person. Where people used to go to coupes to get the high style and performance attributes, they're still going to that mode, but to cars that have four doors.

The Seville and Eldorado share quite a bit of technical content, but they're totally different in looks. They each have a different appeal in terms of the people they were designed to reach.

Author: Do you have a final word?
JG: Definitely, Eldorado is American. Eldorado reflects American engineering, styling, and technology levels that meet and exceed the expectations of this market. And, it is a car that is purchased in the export markets—primarily for Europe and Japan—because of that fact, not in spite of it.

It is recognized as a truly American luxury coupe with a very powerful heritage. The Eldorado will continue to represent the top level of technology for that segment of our business as we move forward through the end of the decade.

Seville is an international car. It's also American. It doesn't have the real American flavor that an Eldorado does. The Eldorado is recognized as American and Cadillac anywhere it's sold around the world, including North America, whereas Seville comes off more as an international car.

I think that's good for us and it's good for the customer. These two cars give us two distinctly different products to present in the marketplace—and they give the customer a choice.

Cadillac Eldorado future concept rendering by designer Scott Wassell. This is still a different way of handling the problem of saying "Cadillac Eldorado" in a design while producing something futuristic that is instantly recognizable as Eldorado. Somehow, the horizontal grooved trim running the length of the car pulls the whole concept together, while the wreathed Cadillac crest on the sail panel and the vertical taillights are just the right nod to the tradition of a great marque.

Cadillac Eldorado future concept rendering by designer Scott Wassell. In this drawing, the same designer who presented the previous rendering attacks the problem in a completely different way. Here Eldorado is shown with a handsome aggressive nose, reminiscent of some 1950s Ferraris, that blatantly challenges the road ahead, while the horizontal body line swoops in an upward angle along the side to define the placement of the rear bumper and taillights. Seemingly smaller than the previous concept, this idea probably more closely approaches the ideas stated by the current Chief Designer Dennis Little while at the same time carrying on some of the same basic design themes of the 1992 design in a bold way.

Cadillac Eldorado future concept rendering by designer Joe Ponce. Also conforming to Chief Designer Little's ideas, this masterful rendering illustrates an automobile that is fairly close to the current model in some ways—for instance, its horizontal side body line, rear taillights, and basic shape of the trunk lid and sail panel. However, this is a spectacularly creative and fresh approach, showing a car of reduced size that still maintains distinctive Eldorado design cues, and at the same time gives an elegant, but sporty, appearance.

Index

Adams, Dan, 33, 48–58, 80, 101
American Sunroof Corporation, 57, 84, 120
Biondo, Frank, 39
Brockstein, Jerry, 80
Buick LeSabre, 15, 17
Buick XP-300, 16
Buick Y-Job, 19
Carli, Renzo, 36
Carter, Julian, W., 134
Casillo, Leonard, 145–149
Cheyne, Charlie, 16, 56
Curtice, Harlow, 49
Dali, Salvador, 49, 50
Doll, John, 48
Donner, Frederick, 87
Dorn, Robert C., 152, 154
Earl, Harley, 5, 7–9, 11, 17, 21–23, 25, 30, 34, 37, 39, 41, 43, 63, 92
Elges, George, 56
Erickson, Larry, 137, 155
Fisher, Lawrence, Jr., 8
Fisher, Marv, 23, 139, 150, 151
General Motors Motorama, 16, 23–25
Glowacke, Ed, 5, 17, 19, 21– 23, 37, 39, 43
Gordon, Jack, 87
Grettenberger, John, 133, 138, 142, 157, 158
Hare, Drew, 119
Hershey, Franklin, Q., 10–12
Hill, Ron, 22, 31, 39, 41–47
Holls, Dave, 17, 21–24, 27, 37, 42, 127, 143
Hopkins, Ted, 122
INFORA Top mechanism, 97
Jordan, Chuck, 37, 47, 67, 75, 80, 82–84, 133, 137
Kady, Wayne, 73, 80, 85, 116–119, 133
Kaptur, Vince, 91–97
Kenard, Edward, C., 111
LaRue, Tom, 80
Lee, Don, 8
Little, Dennis, 156
Lockheed P-38 Lighting, 11
Lutz, Bob, 23
Mark II Continental, 23, 30
McDonald, Jim, 64
Medley, Dwayne, 34
Meyer, Henry, 128–132
Mitchell, Bill, 5, 9, 10, 12, 23, 37, 41, 47, 48, 63, 87, 98, 101, 118–120, 125
Moon, George, 80, 90, 120
Nacker, Owen, M., 8
Nickles, Ned, 5
Northstar System, 129, 131, 132, 152
Ollier, Pierre, 101
Parker, Stan, 73, 80, 85–88, 116
Pininfarina, 36–38, 67, 79, 144, 149, 158
Pininfarina, Battista, 36
Pininfarina, Sergio, 4, 22, 23, 36
Ponce, Joe, 159
Presley, Elvis, 81
Ruzzin, Dick, 136–144
Rybicki, Irv, 63–65, 146
Ryder, George, 22
Saarinen, Eero, 42
Scheelk, Bob, 5, 34, 37, 38–40
Selznack, John, 139
Sitarsky, Wally, 101

Sloan, Alfred, P., 7, 8
Stevens, Brooks, 81, 85
Stevens, Tom, 152
Templin, Bob, 63, 121
Warner, Harold, 87
Wassell, Scott, 158, 159
Wilen, Stan, 120–125

Models
Show Cars
    El Camino, 23, 25, 27
    Eldorado Brougham Town Car 36, 37
    Firebird III, 67
    Jacqueline 4, 53, 67
    La Espada, 26, 27
    LeMans, 16, 17, 23, 26, 63
    LeSabre 16, 17, 19
    Orleans, 16, 17, 25
    Park Avenue, 26, 27
    V-16 Aerodynamic Coupe 5, 8
    Y-Job 19
Eldorado
    1953, 18–20
    1954, 21, 22, 26, 27, 38, 48
    1955, 22, 27–29, 35, 38
    1956, 29, 30, 36–38
    1957, 22–24, 29, 30–34, 36, 38, 39, 40–42, 44, 45, 47, 49, 50, 52, 93
    1958, 22, 23, 29, 33, 34, 36, 38, 46, 49, 50
    1959, 22, 36–38, 48, 51, 52, 55–62, 64
    1960, 36–38, 48, 52, 65, 66, 85
    1961, 67–69, 82
    1962, 43, 47, 67, 68, 70, 71
    1963, 68, 72, 73
    1964, 68, 70
    1965, 73–75
    1966, 80, 81
    1967, 53, 54, 80–83, 85–89, 90, 92, 93, 97, 116, 155, 156
    1968, 93, 94
    1969, 95
    1970, 96
    1971, 55, 56, 99–106, 118
    1972, 104, 107, 108
    1973, 105, 109
    1974, 108, 110
    1975, 108, 110, 120, 125
    1976, 57, 108, 109–113
    1977, 113
    1978, 114
    1979, 83, 97, 114, 115, 117–119, 121–123
    1980, 115, 123
    1981, 115, 124
    1982, 115, 124
    1983, 1151980, 115, 126, 128, 129
    1984, 115, 130
    1985, 115, 128, 131
    1986, 119, 127, 128, 133, 137, 138
    1987, 140
    1988, 140
    1990, 142
    1992, 129–131, 134, 135, 143, 147, 150, 151
    1993, 151, 153
    1995, 154